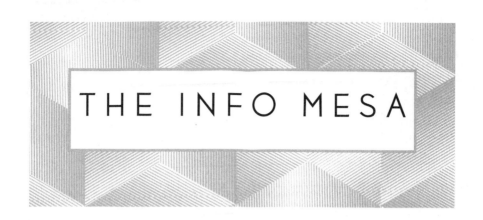

THE INFO MESA

THE

INFO

MESA

SCIENCE, BUSINESS, AND NEW AGE
ALCHEMY ON THE SANTA FE PLATEAU

ED REGIS

W. W. NORTON & COMPANY

New York   London

Copyright © 2003 by Ed Regis

For information about permission to reproduce selections from this book, write to
Permissions, W. W. Norton & Company, Inc., 500 Fifth Avenue, New York, NY 10110

Manufacturing by Quebecor Fairfield
Book design by JAM Design
Production manager: Julia Druskin

Library of Congress Cataloging-in-Publication Data

Regis, Edward, 1944–
The Info Mesa : science, business, and new age alchemy on the Santa Fe Plateau /
by Ed Regis.—1st ed.
p. cm.
Includes bibliographical references and index.
ISBN 0-393-02123-8 (hardcover)
1. High technology industries—New Mexico. 2. Computer industry—New Mexico.
3. Biotechnology industries—New Mexico. 4. Entrepreneurship—New Mexico.
5. Executives—New Mexico. 6. New business enterprises—New Mexico.
7. New products—New Mexico. 8. Science and industry—New Mexico. I. Title.
HC110.T4R43 2003
338.09789'56—dc21

2002155521

W. W. Norton & Company, Inc., 500 Fifth Avenue, New York, N.Y. 10110
www.wwnorton.com

W. W. Norton & Company Ltd., Castle House, 75/76 Wells Street, London W1T 3QT

1 2 3 4 5 6 7 8 9 0

FOR PAMELA

CONTENTS

CAST OF CHARACTERS

In order of appearance.

Dave Weininger   Inventor of SMILES, a chemical nomenclature system. Founding partner, Daylight Chemical Information Systems, Santa Fe.

Dawn Abriel, D.O.   Emergency room physician. Woodworker. Poet.

Stuart Kauffman, M.D.   Complexity theorist. MacArthur Foundation fellow. Professor, Santa Fe Institute. Founding general partner, the BiosGroup, Santa Fe.

Anthony Rippo, M.D.   Physician-entrepreneur. Violin maker. Cofounder, Bioreason, Santa Fe.

Anthony Nicholls   Biophysicist. Founder and "El Presidente," OpenEye Scientific Software, Santa Fe.

George Cowan   Physical chemist. Director for research, senior laboratory fellow, Los Alamos National Laboratory. Founder, Santa Fe Institute.

Enrico Fermi   Nobel laureate in physics, 1938. Manhattan Project scientist, Los Alamos.

Edward Teller   Physicist. Manhattan Project scientist, Los Alamos. Coinventor of the first operational hydrogen-fusion thermonuclear weapon (H-bomb).

Otto Frisch   Physicist. Manhattan Project scientist, Los Alamos.

John von Neumann   Mathematical physicist. Manhattan Project scientist, Los Alamos. Pioneer of the electronic digital computer and of numerical simulation methods and technology.

Emil Hermann Fischer   Nobel laureate in chemistry, 1902.

François Jacob and *Jacques Monod*   Nobel laureates in physiology or medicine, 1965 (with André Lwoff), for explaining genetic regulatory mechanisms in cells.

Warren McCulloch, M.D.   Neurophysiologist, computer scientist, pioneer of neural network theory.

Murray Gell-Mann   Nobel laureate in physics, 1969. Founding board chairman, Santa Fe Institute.

Antoine Lavoisier   Chemist. Inventor of standard chemical nomenclature. Guillotined May 8, 1794.

Jöns Jakob Berzelius   Chemist. Inventor of standard chemical symbolism.

William J. Wiswesser   Chemist. Inventor, Wiswesser Line Notation (WLN).

Harold J. Morowitz   Biologist. Leader, Matrix of Biological Knowledge Workshop, 1987, Santa Fe.

Barry Honig   Biochemist. Creator of DelPhi, program for calculating the electrostatic force fields of macromolecules.

J. Doyne Farmer   Physicist. Cofounder, Prediction Company, Santa Fe.

Norman Packard   Physicist. Cofounder, Prediction Company, Santa Fe.

Marc Ballivet   University of Geneva biologist. Coinventor, with Stuart Kauffman, of a variety of patented processes in biotechnology.

Claude Shannon   Mathematician. Information theorist.

Al Leo   Chemist. Cofounder, Pomona College Medicinal Chemistry Project (MedChem).

Yosi Taitz   Computer scientist with an M.B.A. Founding partner, Daylight Chemical Information Systems, Santa Fe.

Paul Alan Cox   Botanist. Discoverer of prostratin, a potential AIDS drug.

John Elling   Analytical chemist. Cofounder, Bioreason, Santa Fe.

Susan Bassett   Biophysicist and computer scientist. Cofounder, Bioreason, Santa Fe.

John Holland   Computer scientist. Santa Fe Institute scientist. Inventor, genetic algorithms.

Chris Langton   Computer scientist. Santa Fe Institute scientist. Researcher, artificial life. Cofounder and chief science officer, Swarm Corporation, Santa Fe.

Sewall Wright   Evolutionary biologist. Inventor, fitness landscapes.

Laurence Wood   Independent scientist.

Chris Meyer   Director, Center for Business Innovation, Ernst & Young, Boston.

Ruth Nutt   Medicinal chemist. Senior vice president, Bioreason, Santa Fe.

Leroy Hood, M.D.   Biologist. Coinventor, protein sequencer, DNA sequencer ("gene machine"), and other biotechnology devices. Cofounder, Applied Biosystems.

Loralee Makela   Feng shui consultant. Owner, Makela Feng Shui, Santa Fe.

William Ellis   Chemist. Chief, Department of Chemical Information, Division of Experimental Therapeutics, Walter Reed Army Institute of Research.

PROLOGUE

DAVE WEININGER LEVELED off at seventeen thousand feet, sat back in the pilot's seat, and contemplated his situation in the universe. Weininger was the founder, president, and chief scientific officer of Daylight Chemical Information Systems of Santa Fe, New Mexico, and the plane he was flying, a British military jet, was his gift to himself for having invented a new language for chemistry. The language was called SMILES, any given expression within it was also called a SMILES, and that's what Dave had on his face right now, a smile of record-breaking proportions—he was smiling so hard that his cheeks hurt. As he flashed through space at Mach 0.6, six-tenths the speed of sound, high above the Santa Fe mesa, he considered himself the happiest man on the face of the earth.

And why not? On a first-glance view of the matter, inventing a new system of chemical nomenclature was an innovation for which there was no pressing need, an accomplishment on a par with renaming the planets or the stars, or renumbering the galaxies of the Messier catalogue. Indeed, the history of chemistry was littered with failed nomenclatures, none of which had returned lavish financial rewards to their inventors. Dave Weininger's case was quite the opposite, however, for his SMILES nomenclature had made him rich. He had invented a useful product, built an entire company around it, reaped a small fortune in the process, and had rewarded himself with this grand prize.

The aircraft, which he had named *Puer Exuberans*, was a T5A Jet

Provost, a British design, and it was the flying love of Dave Weininger's life. The plane had been intended as a trainer for bombing missions, and the particular model he was piloting had been manufactured in 1969 by British Aerospace. Although his plane had not been fitted out with armaments, the craft had slots along its wings for the placement of two 0.303-caliber machine guns, and on the underside it had weapons racks for the installation of twelve 25-pound rockets as well as two 100-pound bombs. Forget about "handguns," forget about "gun control," Dave Weininger was the proud possessor of his own personal jet bomber.

The blue sky arced overhead, the sun was just rising over the mountains to the east, and the jet exhaust streamed out in an explosive rush, audible in the cockpit as a smooth high whine.

Weininger now visualized the sequence of the aerobatic maneuvers he was about to perform. His routine was written on an index card that was set into the instrument panel in front of him, the whole series of stunts diagrammed in full. But he mentally rehearsed them anyway, the better to make for a flawless performance. Weininger tightened his grip on the control column between his knees and then applied just the slightest amount of back pressure, the same kind of minimalist jolt that an expert rider might apply to a horse's flanks to get it moving from a walk to a trot.

The plane nosed up and soared higher into the blue. It kept rearing back until it was pointed vertically up, and then it went back still more until it was inverted, flying on its back, the brown desert below now appearing directly above Weininger's head. Precisely at that moment, he nudged the stick slightly to the right and the plane responded by executing a half roll, bringing itself right side up, now flying straight and level again, although heading in the opposite direction from before. The maneuver was called an *Immelmann turn*, invented long ago by the World War I German flying ace Max Immelmann, who had used it repeatedly to elude the bullets of pursuing Frenchmen.

Weininger did a few other stunts—a succession of rolls, loops, and a convoluted sequence called a *Cuban-eight*—before calling it quits and heading for home. He never stayed in the air very long while flying his T5A Jet Provost. For one thing, the plane burned fuel at an alarming rate: it had an average fuel consumption of two hundred gallons of Jet A per hour, and at the price of more than two dollars a gallon, this was expensive even for him. And for another, there was the wee matter of landing to consider. Santa Fe Municipal Airport had been designed mostly for commuter planes, small two-engine turboprops and the like. At a length of 8,342 feet, its main landing strip, Runway 2/20, was just barely long enough to accommodate high-speed military jets. Worse, the field was more than a mile above sea level, at an elevation of 6,348 feet where the air was thin, making for a higher-than-normal ground speed at touchdown. Decidedly, the one thing you did not want to do on landing was run out of pavement, and it was Weininger's practice to compensate for the thinner air by landing in the cool of the morning when the air was thicker than it would be later in the day, in the heat of the sun.

He landed shortly after sunrise, taxied to the ramp, shut down the engine, and slid back the electrically operated canopy. He backed the craft into its hangar, closed the doors, and drove home in his Jeep.

Weininger and his longtime mate, Dawn Abriel, an emergency room physician, lived on Stagecoach Road on a ridge north of the city, in a house that had previously been owned by the science fiction and fantasy writer Roger Zelazny. They had purchased the large, sprawling ranch house in 1995, soon after the writer's death. Abriel, whose hobby was woodworking, had replaced the floors in several of the rooms, built a set of elaborate wood shelves in the kitchen, and installed a sink in the baby grand piano that was positioned in the open space between the kitchen and the living room. This was a genuine working sink, complete with hot and cold running water and a drainpipe that went down through the piano bottom and disappeared into the floor below. It was also a genuine working piano, its strings periodically adjusted by a professional tuner who never failed to be amused by the waterworks within.

Weininger's study was off on one side of the house, in the room in which Zelazny had written many of his novels and short stories. A science fiction fan himself, Weininger owned thousands of paperbacks, including some by Zelazny, and had arranged them alphabetically by author on shelves along the room's perimeter. The rest of the space was taken up by computers, disk drives, monitors, and printers—the hardware on which Weininger developed the systems that kept his company afloat, kept it, in fact, awash in cash. SMILES was only one of the many software modules offered by Daylight Chemical Information Systems. Others included SMIRKS, SMARTS, CHUCKLES, CHORTLES, MIRTH, and GRINS, in addition to several others that were not so smilingly named.

The modest goal of all these software systems, and of Dave Weininger himself, was to reduce the world's chemical compounds to information, and chemistry to an information science. He wanted to amass, tabulate, and make available to the user literally all the chemical information in the world. And he wanted to create scientific software of such enormous power, accuracy, and all-inclusiveness that chemists could for the most part bypass the "wet" lab and perform their experiments—the lion's share of them anyway—directly on the computer. Wet labs, after all, took up substantial amounts of space, and running them required money, equipment, and skilled personnel, not to mention ample stores of chemicals, and the experiments themselves consumed large amounts of time and often had to be repeated once, twice, or more.

But at the dawn of the twenty-first century, traditional glass-and-flask chemistry was on the verge of becoming a stone-age relic, not at all attuned to the era in which practically every science, including chemistry, biology, physics, and mathematics, could be and often was reduced to the lightning flow of data through a computer. In the time it took a traditional wet-lab chemist to perform a single experiment with apparatus and compounds, a properly trained computational chemist could perform hundreds, thousands, perhaps even millions of chemical experiments "in silico," yielding results that were just as faithful to the facts as their test-tube counterparts, if not more so.

Conceivably, a properly integrated chemical information system could revolutionize chemistry, agriculture, materials science, and most important of all, drug discovery. Drugs, after all, were the minimally invasive therapy; as expensive as some of them were while still protected by exclusive patents and licensing agreements, pharmaceuticals were far less costly, financially and emotionally, than alternative interventions such as surgery, even microsurgery. So if new drugs could be found that could replace more invasive and expensive treatments, well, the advantages were almost too numerous and obvious to mention. But whereas traditional drug discovery was a long and arduous process, discovering new substances by means of computer experiments was quick and easy. Instead of spending years finding a potential new drug compound in the traditional way, by hand, using old-fashioned laboratory methods, pharmaceutical companies using computational-chemistry techniques could identify promising new compounds in a matter of weeks, days, or maybe even hours.

Weininger's latest project in pursuit of that goal was to bring a hitherto foreign medical tradition into the purview of Western science. He wanted to take the accumulated wisdom of classical Chinese medicine, reduce it to information, express its various compounds in his SMILES language, load it all into a computer-searchable database, and sell the product to modern drug developers, who would use it to create a range of new medical compounds. To aid him in this effort Dave had been teaching himself Mandarin Chinese and had so far learned some four thousand different Han characters.

So after he arrived home from his early-morning aerobatic practice run over the desert, Weininger brewed himself a pot of coffee and poured about a pint of it into a white mug. The mug bore the inscription "MUG '95," which stood for MedChem User's Group and was a souvenir of the annual meetings that Dave's company had been hosting for the past fifteen years. The cup also had a square black panel on the side, and as the heat from the coffee penetrated through to the exterior, the black panel turned into the image of a computer screen covered with color pictures of molecular structures, formulas, and symbols. As the coffee cooled down or was drunk, the on-screen images disappeared—another neat bit of Dave Weininger alchemical trickery.

Mug in hand, Dave disappeared into his study. It was cool and calm in there, and it was also quite dark, for he had painted black over the windows so that he could hardly tell whether it was day or night. So much did he hate glare when he was seated in front of a computer monitor that Weininger even wore a black shirt so that his body produced no reflection in the display screen.

He sat down in front of the large, square monitor and clicked on the TCM icon (the Traditional Chinese Medicine application). A window opened up showing a list of Chinese ideograms plus a text box. He typed the Chinese words *ci wu jia* into the text box, pressed <ENTER>, and there appeared on the screen a list of numbered medical molecules, their Chinese names, their Western equivalents, and their SMILES, together with a color-coded structural diagram for each separate chemical.

He worked on the program for the rest of the day.

DAVE WEININGER, HOWEVER, was not the only scientist in Santa Fe pursuing the task of converting external reality into files of abstract scientific information and making large amounts of money at it. A mile away downtown in a long, low building on Paseo de Peralta, one Stuart Kauffman, M.D., was doing much the same.

The founding general partner and chief scientist at the BiosGroup, Kauffman was one of the iconic figures in Santa Fe science. First and foremost, he was a founding father of complexity theory, an overarching interpretive schema that was intended to embrace and explain virtually all of nature. The central tenet of complexity science—that complex wholes resulted from the repeated interactions among the simpler elements of which they were composed—was presumed to explain phenomena as diverse as the origin of life on earth, the development of a single fertilized cell into a complex and differentiated organism, the food-foraging strategies of ant colonies, the genetics of fruit flies, the workings of the immune system, and the existence of order in the universe at large, among other things. Kauffman, it turned out, had pursued fundamental research on all of these subjects and had published long, technical treatises about them. In July 1987, in recognition of his breakthrough work in developmental and evolutionary biology, Kauffman was awarded a five-year John D. and Catherine T. MacArthur Fellowship, the so-called genius grant, which at that time was worth $290,000.

But in fact, theorizing about such abstract issues was only one small part of Kauffman's overall intellectual cosmos. Kauffman had an entirely separate, practical side that was reflected in the nine patents he held on inventions having to do with the large-scale production of peptides, polypeptides, and proteins. Besides this, Stu Kauffman was a physician; he had even practiced hands-on medicine for a short time, although he realized early in the game that the healing arts were really not his true métier. As to what was, however, that was more difficult to say, the reason being that he had a passion for so many things—and making money was not the least of them. He was not invariably successful in that effort: in 1992 he had founded a company called Darwin Molecular to capitalize on his patents, but was soon banished from the firm by the other partners.

Still, it was somewhat of a surprise even in the face of all this that in 1996, when he started a second company, the BiosGroup, the firm, despite its title, had nothing whatsoever to do with biology, strictly considered. The BiosGroup's corporate motto was "Science for business," and the company would apply complexity theory to some of the knottier problems of modern commerce. Soon Kauffman and a scientific staff that eventually reached to over a hundred were solving otherwise intractable business problems for a collection of Fortune 500 companies including Procter & Gamble, Southwest Airlines, Ford, Boeing, Texas Instruments, and the Walt Disney Company.

How a nonpracticing physician who was one of the creators of complexity theory, who held patents on novel methods of oligonucleotide production, and who for more than twenty years had been a research scientist of the purest stripe had apparently junked all that in one fell swoop and then waded into grimy industrial waters where he was pulling in vast amounts of cold cash was something of a mystery.

As was the fact that practically overnight, Santa Fe had turned into a hotbed of science startups that were similarly intent on making fortunes from some of the most advanced and abstruse science imaginable. Why here? Why now? And why just these people?

There was Bioreason, for example, whose chief executive officer, Anthony Rippo, himself a nonpracticing physician, violin maker, and entrepreneur, wanted to automate drug discovery to the point that live human beings were substantially excluded from the process, both as experimenters and as guinea pigs. Instead, automated reasoning programs—the mental processes of the world's best drug discoverers effectively downloaded into computers—would find the new drugs and assess their potential efficacy against a range of human and animal diseases, sparing medicinal chemists and lab workers the time and trouble. Another firm, PHASE-1 Molecular Toxicology, would tell pharmaceutical companies which, if any, of their most promising drug candidates might be toxic to humans or animals.

Flow Science would solve any problem in fluid flow with its FLOW-3D software. Another company, Complexica, had developed a complexity theory–based program for the insurance business. The application, called Insurance World, simulated the entire global insurance industry and would tell the user where, why, and how grave the worst risks were.

By 2002, indeed, a whole fleet of science-applications companies had arisen and were flourishing on the Santa Fe mesa: Adaptive Network

Solutions Research, DNA Mining Informatics Software, eOrder$ource, Genzyme Genetics, Metaphorics, Molecular Informatics, the National Center for Genome Resources, QTL Biosystems, the Prediction Company, Strategic Analytics, and the Swarm Corporation, more than two-dozen recent startups, all of which specialized in applying one or more nouveau scientific theories—complexity theory, chaos theory, nonlinear dynamics, artificial life—to problems in the real world and (they hoped) making a profit from it.

It got to the point where Santa Fe, which the scientist-entrepreneurs there took to calling "the Info Mesa" for its success at turning raw data into information (and then dollars), was looking ever more like a new species of Silicon Valley, a place where science and technology firms gathered together, did business, and created new levels of wealth in the process. The Info Mesa even had its own unofficial Steve Jobs figurehead in the person of Anthony Nicholls, founder, chief scientist, and as he branded himself, "El Presidente" of OpenEye Scientific Software. Working out of his tiny adobe apartment located on a dirt backstreet in Santa Fe, Nicholls was writing a suite of new-wave protein-visualization programs and other types of small-molecule-chemistry applications that he hoped would blow away researchers in chemical, pharmaceutical, and agricultural industries.

All of this was quite anomalous. How did it happen that a smallish and remote town at the foot of the Sangre de Cristo Mountains, the oldest continuously inhabited city in America, home of New Age mysticism, more than a hundred art galleries, and "Santa Fe style," and whose principal industries in all previous eras were tourism, turquoise, and Indians, had suddenly remade itself as a hotbed of recherché science, advanced technology, and money, a nexus that would transform the hitherto accepted way in which scientific innovation was pursued and funded?

There was, of course, an answer to the question, but the chief clues were not to be found in Santa Fe. In fact, the defining moment in the story of the Info Mesa occurred not in the city itself but rather on a small, sunlit island in the central Pacific, where in the fall of 1952 a young radio-chemist by the name of George Cowan was checking up on the health of his two fast-neutron detectors.

PART ONE

THEORY

THE LOS ALAMOS PROBLEM

ELUGELAB WAS A small, diamond-shaped island located eleven degrees above the equator some three thousand miles west of Hawaii. The island was part of an oval-shaped coral atoll consisting of forty-four separate islands that enclosed a large open lagoon. The atoll was quite remote, the nearest substantial land mass being Papua New Guinea, almost a thousand miles away to the south.

Elugelab was covered with tropical grasses, wildflowers, shrubs, and small trees known as *Messerschmidia*, but possessed no human inhabitants or mammals of any other sort. Nevertheless, several species of tropical birds could be found on its outer beaches, which were mostly coral rock and gravel, and on the interior of the island, which was largely smooth white sand. There on any given day hundreds of white terns and sooty terns, brown noddies, and wedge-tailed shearwaters were nesting, sleeping, or foraging around through the ground cover. Just before sunset, golden plovers or frigate birds would fly in stately fashion across the lagoon, often skimming only inches above the water.

In March 1952, an American military task force arrived on this primeval, forty-acre tropical outpost and proceeded to remake it from the ground up, for the island had suddenly acquired a starring role in human affairs. Bulldozers rolled onto the shore from landing craft and piled up the sand and coral so as to create an elevated flat spot in the center. On it, construction crews erected a six-story-high test structure called the *shot cab*, and behind it, a 375-foot-tall radio and television mast.

Running in a straight line northeast from the shot cab there appeared a nine-thousand-foot-long aluminum-sheathed plywood tunnel. The tunnel, which looked like a long covered bridge, was wide enough for four or five men to walk through abreast. The bridge was called the *helium box* by those who built it because it would eventually be filled with helium gas (it was also known as the *Krause-Ogle box,* after its inventors). The box stretched across four separate islands: from Elugelab along a manmade sand and coral causeway to the neighboring island of Teiter, then across a second causeway to tiny Bogairikk, and finally over a third to Bogon, a larger, spade-shaped island almost two miles away from the first. The sides of the tunnel, as well as the roof and the floor, were so rectilinear and true that the curvature of the earth below it caused both ends to be suspended slightly above ground level, like a balance beam atop a large ball.

The excuse for all this construction activity was another device called the *Sausage.* The Sausage was the world's first thermonuclear weapon, a hydrogen bomb, and it was sequestered inside the shot cab much as if it were a new species of secret rocket hidden away inside a hangar.

The bomb was to be detonated on Elugelab before the year was out. If the test was successful and the device exploded on schedule, then gamma and neutron particles from the fusion reaction would stream through the helium box to data-collecting instruments at the Bogon end, which was called *Station 200.* From Station 200 the incoming data would be telemetered to receivers farther away while the shot cab, the helium box, the measuring instruments, and everything else in the immediate vicinity disappeared in a blinding flash of heat and light.

That, anyway, was the plan. At this point, however, considerable uncertainty was still attached to the outcome. For one thing, the device inside the shot cab was the very first specimen of its kind, and it was based on a novel, untested, and wholly unproved design. It was an open question, therefore, whether the apparatus as it stood would work at all.

Second was the fact that in view of the nature of the device, its mode of ignition, and the probable consequences of a successful firing, there had been no question of doing any initial, exploratory, proof-of-concept tests using a smaller, scaled-down, laboratory version of the bomb. The H-bomb was a fusion device, meaning that the explosion inside it would be the product of the nuclei of lighter elements (such as deuterium, an isotope of hydrogen) fusing together to form nuclei of heavier elements (such as helium), releasing immense quantities of energy in the process. The problem was to get the fusion reaction to begin and, once begun, to keep it going without fizzling out. Because atomic nuclei by nature didn't want to fuse together on their own, it would take extremely high levels of heat energy to force them to do so. At Los Alamos, the home of American atomic bomb research, the idea for generating the necessary energy was to use a "conventional" atomic bomb (a fission bomb) as a starter device, a sort of spark plug that would induce the fusion reaction. In other words, an ordinary A-bomb, the type of device that had been tested at Trinity Site and then detonated over Hiroshima and Nagasaki, would ignite the fusion reaction inside the H-bomb, which would respond by exploding in the grand manner.

The idea of using a fission bomb to set off a hydrogen bomb had come to Enrico Fermi one day in September 1941 as he and Edward Teller were walking across the Columbia University campus, where they were both professors of physics. For no particular reason, Fermi suddenly blurted out to Teller the notion that the heat of an atomic blast might be sufficient to ignite a fusion reaction in a mass of deuterium, releasing enormous quantities of destructive energy in the process.

"I thought about that question for a week or so," Teller said later. "And then, during one beautiful autumn Sunday afternoon walk with Fermi, I explained why the reaction wouldn't work: At the very high temperatures we expected, most of the energy would not be present as light but as x-rays. X-rays would be radiated away without bringing the nuclei and particles closer together; they were useless for carrying on the reaction."

That is to say, the necessary temperatures would be attained momentarily, but the fusion reaction would die out (if it even got going) because the heat energy would be radiated away faster than it could be utilized to force the deuterium nuclei together.

"And I explained to Enrico why a hydrogen bomb could never be made," Teller said. "And he believed me."

Two years later, though, everything had changed. Fermi and Teller were now working on the fission bomb project at Los Alamos, the secret mountaintop lab about thirty miles northwest of Santa Fe, and in the interim Teller had changed his mind about the hydrogen bomb, which he called "the Super." He now thought that the Super could work, and that Fermi's idea for setting it off with the heat from a fissioning atomic bomb was a possible means of doing so.

"We found an objection to my earlier objection," Teller said. "The x-rays that I had assumed would siphon off the energy are not emitted at once. Perhaps the reaction could proceed before too much of the energy was lost to x-rays."

Well, perhaps. Still, this was just one more scientific speculation, a hypothesis, a guess that required experimental verification or disproof. The question was, how to decide the issue experimentally, one way or the other? You couldn't set off a little backyard, firecracker version of the A-bomb to see if it would successfully ignite a little backyard, firecracker version of the H-bomb. There just weren't any miniature desktop examples of these bombs available, nor was there any apparent way of creating one that was both small enough to constitute a genuine scale model and large enough to ignite and sustain the fusion reaction. If the triggering device was below a certain threshold of size and mass, then the heat it generated would be too feeble to start the required course of nuclear fusion, much less propagate it through to the end.

No, if you were going to ignite devices of this kind at all, they had to be of a certain minimum level of thermal energy, power, and destructiveness, and if a small Pacific island had to be sacrificed in the process, well, such was the price of progress in H-bomb research.

Matters had been different with the atomic bomb itself, which worked on the principle of nuclear fission. On very small scales, and for very brief periods, controlled fission had been achieved many times in the laboratory. The first occasion had been in December of 1942, when Enrico Fermi produced a controlled atomic chain reaction in an underground room that had once been a squash court at the University of Chicago. That was a historic event, regarded ever since as the birth of the atomic age.

Then there were the famous "tickling the dragon" experiments at Los Alamos, designed to settle the question of exactly how much uranium would be required for a workable fission bomb. Theory gave an answer to the question, but engineers knew from long and bitter experience that there was often a sizable mismatch between what theory claimed and what practice revealed to be the truth. It was therefore desirable to have the theoretical answer verified by empirical tests, but it was not obvious how to go about this safely. In 1941, Otto Frisch, a Los Alamos physicist who had escaped from Nazi Germany (and who, together with his aunt, Lise Meitner, had proposed the term *fission* to refer to the splitting up of an atomic nucleus), now came up with the idea of physically assembling together the quantity of uranium-235 (U-235) that theoretical calculations said would be just shy of an explosive amount, and then briefly raising the quantity to see if it exhibited signs of "going critical," which was to say, exploding.

The centerpiece of Frisch's experimental apparatus was the "guillotine," a ten-foot-tall steel frame that would drop a small slug of dilute U-235 through a hole in a pile of subcritical uranium. The slug was about the size and shape of a fat cigar, and it would bring the pile to criticality for only the briefest instant, but for long enough to know whether the experiment had succeeded or not. "It was as near as we could possibly go towards starting an atomic explosion without actually being blown up," said Frisch much later. Indeed, so risky was the scheme that Richard Feynman, who also worked on the device, said the procedure would be like "tickling the tail of a sleeping dragon."

Nevertheless, the Los Alamos scientists built the contraption in a small wood-frame building in a remote canyon, assembled the necessary U-235, and then dropped the slug down through it. "The results were most satisfactory," Frisch reported. "Everything happened exactly as it should. When the core was dropped through the hole we got a large burst of neutrons and a temperature rise of several degrees in that very short split second during which the chain reaction proceeded as a sort of stifled explosion."

Still, those experiments concerned *fission*, and there was no precise analog to them that would work for fusion. To unite atomic nuclei against the repulsive forces that operated to keep them apart, extremely high and reasonably sustained energies were required, temperatures that a fullfledged atomic bomb, and not merely a "stifled explosion," would be needed to generate.

Fortunately, there was a third alternative between theory on the one hand and a full-scale fusion experiment on the other: there was *simulation*. If you could somehow reproduce in another medium the physical process involved in genuine nuclear fusion, then perhaps you could determine beforehand whether the fusion reaction would ignite and propagate through to completion.

A simulation, the atomic scientists hoped, would save the day, telling them what they wanted to know so that they wouldn't find themselves in the embarrassing situation of having designed, built, and tested an inordinately expensive and fabulous dud.

SIMULATIONS WERE A common tool not only in the development of a new technology but also in countless other research contexts in which a cheaper, smaller, surrogate version of a given phenomenon could afford some insight into the operation of the real thing. An astronaut's working underwater simulated the weightlessness of space; aerodynamic scale models in wind tunnels simulated the lift, drag, and turbulence-generating properties of full-size aircraft; small, plastic snap-together spheres simulated atoms and the structure and workings of molecules; and the movements of toy soldiers on a war-games board simulated the progress of troops in battle. All these counterfeit objects were substitute, stand-in versions of the thing itself, whose workings they reproduced in greater or lesser degrees of fidelity.

Stretching a point, even the use of mathematical equations in physics was a weak form of simulation, as the equations represented in another medium (a formal language) relationships that existed in the empirical world. When physicists used the formula $F = ma$, for example, to represent the equivalence between force and accelerated states of matter, that equation was a means of expressing in numerical language the external behavior of matter itself and was, as such, a reproduction, albeit static and lifeless, of the real thing.

When it came to predicting the behavior of the hydrogen bomb, the Los Alamos scientists had plenty of mathematical equations to work with—indeed, they had too many of them. There were equations governing the propagation of shock waves in various media, the transport of electromagnetic radiation through space, rates of neutron emission, and so on and so forth. The problem, however, was that in the case of the hydrogen bomb, those equations were too difficult to solve by traditional analytical methods, especially in view of the fact that all those numerous interconnected processes would be at work simultaneously inside the bomb so that the course of one reaction would be affected by that of another, and so on down a long list. Stanislaw Ulam, a mathematician who worked on the bomb project with Teller, described the complexity of the problem: "All the questions of behavior of the material as it heated and expanded—the changing time rate of the reaction; the hydrodynamics of the motion of the material; and the interaction with the radiation field—had to be formulated and calculated. The problem of the start and explosion of a mass of deuterium combined a considerable number of separate problems. Each of these was of great difficulty in itself, and they were all strongly interconnected."

To predict the behavior of the thermonuclear bomb, researchers therefore resorted to what they called *numerical methods*, which amounted to an incremental, stepwise, moment-to-moment simulation of the relevant physical reactions as they proceeded. Literally, the use of numerical methods meant calculating a solution for the physical state of affairs at one given time-step in the developing reaction, then performing a second round of calculations to establish the situation at the next time-step, another for the third, and so on, one after the next, until the full set of numerical computations had reproduced the entire course of events, or at least a series of successive numerical snapshots of it, over time. Such a sequential, reiterative process in effect reproduced, on paper, an approximation of the reaction as it would actually unfold in the real world. It was like following the descent of a hotel elevator by watching the blinking lights on the annunciator panel on the ground floor.

The Los Alamos scientists had pioneered the use of numerical methods earlier, during the mid-1940s, as they were designing the fission bomb. In that case, the atomic explosion would result from a chain reaction in a critical mass of uranium or plutonium, but it was not clear how best to assemble the mass in question. One idea, the gun method, involved firing a radioactive slug into a subcritical mass of plutonium, causing it to go critical immediately upon the slug's entrance. The problem with such a scheme, however, was that the two parts might predetonate, the bullet and target melting down and the explosion snuffing itself out before it got fully under way.

A second method for producing a chain reaction was by implosion, which meant crushing a hollow shell of individually subcritical pieces until they came together and formed a solid sphere of critical mass, thereby inducing the fission reaction. The implosive forces would be provided by conventional high explosives staggered evenly around the shell's external surface.

Implosion was a highly novel and speculative concept inasmuch as explosives were normally used to blow things apart rather than to blow a collection of disparate pieces together into a smooth and tidy sphere. The virtue of the implosion arrangement, however, was that driving the pieces forcibly inward would place them all together at the same instant and hold them there long enough for the entire mass to go critical simultaneously. The disadvantage was that none of this had been done before, and nobody knew whether, or how well, the implosion scheme would work.

The task facing the Los Alamos scientists, then, was to calculate the successive positions of the implosion shock wave through time so that they could follow the progress of the plutonium shell as it drew together to form a compact, spherical, and explosive mass. With the relevant equations overly complicated even in this case, they responded by finding the answers through numerical methods. This, however, involved tracing the paths of hundreds or thousands of individual particles of matter as they flew together toward a common point, and the calculations required were repetitious in the extreme.

The Los Alamos scientists therefore turned to machines. Machine computation, such as it was, had reached only a primitive state both at the Los Alamos lab and in the outside world. In the spring of 1943, nevertheless, in order to get on with their work, the lab took delivery of a bunch of Marchant and Friden desk calculators. These were simple adding machines powered by hand cranks—manual devices that performed addition, subtraction, multiplication, and with a little coaxing, even division.

The machines were self-contained universes of meshing matter, made up of gears, cams, pushrods, and ratchets, and the dirt and dust that were a constant presence on the Los Alamos plateau caused them to break down on a regular basis, in addition to which their human operators were themselves subject to confusion, fatigue, temper tantrums, and mistakes. Soon enough the mechanical calculators were replaced by something more reliable—a set of IBM punched-card machines. These machines were mechanical devices too, with lots of metal sliding against metal, but at least they operated automatically, without much need for human intervention, and for that reason alone they were a substantial advance over the manual calculators. And although the IBM units had been designed as business machines and were originally intended mainly for accounting purposes, a Columbia University astronomer by the name of Wallace Eckert had used them successfully to calculate astronomical tables that traced the motions of celestial bodies through space. There seemed to be no reason why the same devices couldn't trace equally well the path of an implosion shock wave through matter.

The IBM machines arrived at Los Alamos in the fall of 1943 and were doing implosion computations within a week. Key to the operation was a piece of stiff card stock about three inches by seven; by punching holes in it at various locations, a given card could be made to represent a specific scientific data point. These "IBM cards," later to become items of folklore, had been invented by Herman Hollerith in 1890 to record census data, but they could hold information of any type.

An implosion simulation involved tracing the advance of a three-dimensional shock wave over time, and so the Los Alamos scientists followed the wave's progress by cutting a separate punched card for each point along the wave, such that a stack of IBM cards stood for a cross section of the entire implosion wave at a given instant of time, t_1. Then they ran the cards through the IBM machine, which by calculating ahead on the basis of the relevant equations, estimated the wave's position at the next instant of time, t_2. This resulted in a new deck of punched cards, which were themselves run through the machine to determine the wave's position at t_3, and so on. At each successive time-step, a tabulating machine printed out the numerical values that established the wave's new position in space.

To monitor the computations of the IBM machines, the Los Alamos scientists compared their outputs with the results from the adding machines, whose human operators were put to work on the same problem using the same sets of equations and data points that had been fed into the IBMs. And to motivate the humans to their highest levels of speed and accuracy, the lab chiefs set up a man-machine competition. "For the first few days the hand-computing group kept approximately even with the punched-card machine operation," said Nick Metropolis, who would emerge as the Los Alamos lab's top computer whiz. "But on the third day the punched-card machine operation began to move decisively ahead, as the people performing the hand computing could not sustain their initial fast pace, while the machines did not tire and continued at their steady pace. The race was then abandoned."

When it was all over and done with, the humans working with mechanical calculators agreed with the automated IBM machines that a shell of plutonium would indeed be compressed to a ball of critical mass if the shell were collapsed uniformly by a spherically symmetrical implosion. Two separate sets of numerical simulations, in other words, said that the device would work.

There lay ahead one of the biggest problems with any simulation, which was ensuring its fidelity to the set of events in external reality that the simulation was designed to replicate—the problem of verification. In this instance the required proof was not long in coming, and on July 16, 1945, at Trinity Site in the New Mexico desert near Alamogordo, the events predicted by the numerical simulations were faithfully enacted in the real world when a group of detonators surrounding a plutonium core the size of a cantaloupe fired simultaneously, squeezing the core to about the size of a human eyeball, creating a critical mass, and setting off the chain reaction inside the world's first atomic bomb.

The previous numerical simulations were milestones of machine computation, for they showed that a complicated set of physical conditions could be reduced to abstract numerical data points, that those data points could be mathematically manipulated by machines in such a way as to reflect the actual progress of a complex event in the real world, and that the machine's numerical output could be translated back into usable information that accurately predicted a future state of empirical reality.

It was precisely that ability to reduce empirical reality to data, to manipulate that data by machines, and then to extract from the output an important new empirical result, that years later underlay the birth of the Info Mesa.

STANDING IN THE shade of the shot cab on Elugelab, George Cowan took a long last look at the enclosure that housed his test equipment. Cowan was a radiochemist who had come to the Eniwetok Atoll, of which Elugelab was a part, in the fall of 1952 to record radiation data from the explosion (assuming that it would occur) of the world's first hydrogen bomb. His test instruments were designed to measure the flux of fast-neutron radiation as it streamed out of the bomb core during the fusion reaction. As Edward Teller once described it, "A thermonuclear reaction between deuterium and tritium produces fast neutrons. They leave the neighborhood of the explosion (with a velocity of more than 1,000 miles per second) and can be easily found a few hundred feet away from their origin. Such neutrons will bump into protons, and they will leave characteristic long tracks when captured on a photographic plate. Such a long track is unmistakable evidence of a thermonuclear reaction."

Cowan's fast-neutron detectors were attached to the bomb's external casing by a length of pipe. Once the reaction began and the nuclear fusion was under way, his detectors would have an exceptionally short working life—all of a microsecond, perhaps—but that interval would be long enough for his instruments to record the necessary observations and send the data through the nearly two-mile-long helium box to the transmitting equipment at Bogon, which in turn would radio them to receivers on the USS *Estes*, which was where Cowan would be stationed for the test.

Later, Cowan would become director of research at the Los Alamos National Laboratory. In time, the lab would become one of the world's largest users of computers, supercomputers, and scientific simulations of physical phenomena—simulations devoted chiefly, but not exclusively, to the detonations of thermonuclear bombs.

Later still, after he resigned as research director, Cowan would play a formative role in the birth of the Info Mesa, for it was he who founded the scientific institution that brought complexity theory, chaos theory, and the scientists who pioneered these studies to Santa Fe. This was the Santa Fe Institute, which Cowan created in 1984 to promote a new form of interdisciplinary scientific research. Such research emphasized a new, antireductionistic approach to nature, one that favored a more holistic viewpoint in which the mutual interactions among elementary constituents yielded a range of novel emergent phenomena—structures and behaviors that were not easily predictable from a narrow, reductionistic description of the individual agents themselves.

All that was far in the future. Still, it would be a hallmark of the Santa Fe Institute's new way of doing science that computers, and especially computer modeling of emergent phenomena, were to be used all over the place, all the time, by every scientist who ever worked there. The machines would become iconic commodities, omnipresent at the institute—and indeed practically omniscient or even omnipotent deities. In fact, the Santa Fe Institute's whole approach to the understanding of nature would hardly be possible without them.

On the surface, it was implausible that one of the original atomic scientists would become the intellectual godfather of the Santa Fe Institute, complexity science, and the Info Mesa, but in retrospect all of it made perfect sense. For the Santa Fe Institute, as it turned out, was a sort of magnet and staging point for the Info Mesa, and George Cowan was the middleman, the intermediary, the facilitator, charting a path between simulations of nuclear devices on the one hand and simulations of equally complex emergent phenomena of numerous and wholly different types on the other.

There was a synergistic, symbiotic relationship, after all, between the bomb, the computer, and complexity science. Bomb research drove the invention of bigger and better computers, while the computers in turn made the bombs, as well as complexity theory itself, possible. Without the H-bomb, and without the computer technology research and development that simulated the bomb's inner workings, neither the Santa Fe Institute, complexity science, nor the Info Mesa itself could have existed as and when they came to be. The technology underlying the bomb, the Santa Fe Institute, and the Info Mesa was developed in the seven-year period between 1945, when the first scientific problem run on the world's first electronic digital computer was whether the H-bomb would explode, and 1952, when, in front of George Cowan's very own eyes, it did.

By the time he'd landed on Elugelab in 1952, Cowan was already an old hand at the relatively new business of studying the practical uses of radioactive elements. In 1941, while still in his twenties, he had worked on the Princeton cyclotron project, taking measurements that would determine whether a chain reaction could be produced in uranium. A year later he was at the University of Chicago, where as a junior member of the Metallurgy Lab, he had worked on Enrico Fermi's original atomic pile, the mass of uranium that underwent its historic moment of controlled fission in December 1942.

At that point, well before the first atomic bomb existed as a piece of equipment, and even before the founding of the Los Alamos lab that would design, produce, and test it, the atomic scientists were already thinking about an even more powerful device—the Super.

In the summer of 1942 Robert Oppenheimer had organized a secret conference held in LeConte Hall on the campus of the University of California, Berkeley. Supposedly, the subject of discussion was the fission bomb. Nevertheless, "the discussion that summer wasn't confined to fission," recalled Robert Serber, one of the participants. "We reviewed the theory, but everyone seemed to be saying, well, that's all settled, let's talk about something interesting."

What Edward Teller wanted to talk about was the hydrogen bomb, the Super.

"He started bringing in all kinds of wild ideas," said Serber. "Edward was always full of ideas. The main thing Teller was hooked on, of course, was the idea of pushing through to a thermonuclear weapon, an H-bomb, the so-called classical Super. It was essentially similar to TNT—a detonation wave moving through a deuterium and tritium mixture. Edward raised this question during the summer 1942 conference and got everybody interested."

"We proceeded with some excitement to find out whether we could make a thermonuclear reaction proceed," Teller himself recalled later. "During the next few weeks, we convinced ourselves it could be done."

Teller's Super would be the ultimate weapon. Whereas the Trinity bomb's explosive yield would be on the order of twenty kilotons (equivalent to twenty thousand tons of TNT), the Super, assuming it could be made to work, would have a yield measured in megatons (millions of tons of TNT). Indeed, a relatively average ten-megaton H-bomb would be equal in destructive power to five hundred Trinity-type bombs.

The bomb as Teller first conceived of it would use deuterium as fuel, but there were a couple of problems with deuterium. The first was that temperatures on the order of 400 million degrees would be required to start the reaction. Conceivably, such temperatures could be produced momentarily by a fission bomb, but that only led to the second problem, that of heat dissipation. Temperatures of 400 million degrees, even if briefly attained, would cool down instantly by thermal radiation, snuffing out the fusion reaction almost before it began. The H-bomb problem therefore reduced itself to two separate questions: One, would a fission bomb ignite the fusion reaction? Two, would that reaction, once ignited, persist to completion?

At Los Alamos, where the scientists could talk about the H-bomb freely, these two questions came to be known as "the Super problem." Outside the lab, where other scientists were officially not supposed to know that designs for a Super bomb even existed, it was referred to instead as "the

Los Alamos problem." In the fall of 1945, not long after the Trinity test, the Los Alamos problem was given to the ENIAC, a new species of computer that had just been built in Philadelphia.

The ENIAC (Electronic Numerical Integrator and Computer) was a product of the University of Pennsylvania's Moore School of Engineering. The machine was the world's first electronic digital computer and the original brontosaurus of the species. The ENIAC consisted of 18,000 vacuum tubes, 70,000 resistors, 10,000 capacitors, 6,000 switches, and a half million solder joints, all crammed into forty tall slabs of electronics that together weighed thirty tons. The machine had been designed to calculate shell trajectories and firing tables for the U.S. Army's Ballistic Research Laboratory, a task it would eventually perform with what was then regarded as great speed. (It was considered a triumph of computation when it calculated the trajectory of a sixteen-inch naval projectile in less time than it took the shell to cross the distance.)

In the fall of 1945, the Los Alamos scientists sent 1 million punched cards to the ENIAC team in Philadelphia. The exact physical nature of the problem was classified and could not be told even to those who would run the simulations. The equations involved were unclassified, however, and the numerical data would be stored in the form of punched cards, which were individually and collectively unintelligible, and in that way a measure of secrecy was preserved.

All those who worked on the project were impressed by the size and scope of the problem and by the number of calculations that would be necessary to solve it. Each tried to outdo the other in providing a fitting image for magnitude of the task. John von Neumann, for example, said, "This computation will require more multiplications than have ever been done before by all of humanity." Stanislaw Ulam said, "Ours was the biggest problem ever, vastly larger than any astronomical calculation done to that date."

Today's computers could work the problem in a matter of seconds; back then, the ENIAC took weeks. In fact, the ENIAC worked on the Los Alamos problem over a six-week period between December 1945 and January 1946, all the while printing out endless reams of numerical results. The question was, what did its Delphic utterances portend?

In April 1946 the top atomic scientists held a secret three-day "Super" conference at Los Alamos for the purpose of figuring out what, in fact, the ENIAC had said. Almost four years later, in February 1950, a secret "Report of Conference on the Super" was issued by the Los Alamos lab and distributed to the thirty-one conference participants, including Nick Metropolis, John von Neumann, Stan Ulam, and Edward Teller. In 1971, finally, portions of the report were declassified and made available to researchers.

The final conclusions of the report, based on work done by the ENIAC and confirmed by hand calculations, were as follows: "It is likely that a super-bomb can be constructed and will work," and "Definite proof of this can hardly ever be expected and a final decision can be made only by a test of the completely assembled super-bomb."

A later series of computer calculations, by contrast, said that the device would *not* work. But the scientists soon realized that this was because the computer's time-steps had been too coarse to capture the fine-grained details of the phenomenon. "Today, the shortcomings of those calculations are widely known," said Edward Teller. "Most important is that reality proceeds in infinitesimal time-steps, but a computer operates in finite steps. That difference can be tolerated if the steps are small enough. But the early computers could not accommodate many steps, and, therefore, the steps were comparatively large."

Proof of the pudding, in the end, would have to come from experiment.

Assembly of the H-bomb was a slow and halting process, the reason being that there were still several competing design geometries on the drawing boards, plus new ignition schemes and explosive media, as well as other proposed innovations, all of which had to be simulated on computers. But the Los Alamos scientists and others invented new and faster computers, machines that were sufficient to the task. There was the Institute for Advanced Study computer, built by John von Neumann at Princeton. There was the MANIAC (Mathematical and Numerical Integrator and Computer), built by Nick Metropolis at Los Alamos. There was the SEAC (Standards Electronic Automatic Computer), built at the National Bureau of Standards in Washington.

The inventors again made imposing claims about the size and scale of the computation. The simulation run on the Institute for Advanced Study machine was "the first large problem that was done on it," von Neumann reported. The problem "was quite large, and took even under these conditions half a year."

The H-bomb that finally took shape on the island of Elugelab was by far the most complex and intricate piece of physical apparatus ever constructed until that time. It would operate, if it worked at all, on the basis of a reaction that had been produced only once before on planet Earth. About a year and a half earlier, on May 9, 1951, on another island of the Eniwetok Atoll, there had been a test called *Greenhouse George* (named after its designer, physicist George Gamow). The device in that case had been a pure fission bomb, set on a two-hundred-foot-high tower on the neighboring island of Eleleron. But that bomb had included, as an external appendage, a bit of the deuterium-tritium mixture that would power the Super. When the George bomb exploded, the deuterium-tritium mixture underwent a short burst of fusion that produced a scattering of fast neutrons. It was as close to a laboratory version of the fusion reaction that the atomic scientists ever achieved.

Still, that was only a test device, not a weapon. "George was an experiment," Teller said later. "The device had no military usefulness; the amount of tritium involved would have made it too expensive to be a practical weapon."

The test of the Super would take place in late fall of 1952.

After finishing the final inspection of his fast-neutron devices, George Cowan, along with the few other scientists then remaining on the island, boarded a launch for the steamship *Estes*. The ship now headed for a point thirty miles away from Elugelab in preparation for the test shot, which had been code-named "Ivy Mike" ("Mike" for megaton).

At seven o'clock in the morning of November 1, 1952, Cowan was on the main deck of the *Estes*, watching and waiting as shipboard loudspeakers announced the countdown.

Soon it was 7:14.

The coupling of theory, instrumentation, data, and machine computation had led, for better or for worse, to this precise moment in the history of science and technology. Scientists had seen deeply into the fine structure of matter, had abstracted away its inner essence in the form of numerical data, and had manipulated that data to find out what the next natural steps would be in the lawlike sequence of nuclear events—the structure of reality having been mirrored in the silent flow of electrons inside the first large-scale computing machines.

Less than a minute later, the count reached zero.

2

THE DEEP

IN 1952, THE same year that the world's first hydrogen bomb was detonated at Elugelab, Marion Weininger gave birth to her first son, David, at Woman's General Hospital in New York City. Dave's father, Joseph Leopold Weininger, had escaped from Austria during the war years and had emigrated to the United States, traveling through England and then Canada, where for a while he was interned as an "enemy alien." (Some other Weininger family members, not so fortunate, had perished at Dachau.) After the war Joe Weininger studied chemistry at McGill University in Montreal, which is where he met his future wife.

At the time of Dave's birth, the Weiningers were living on Congress Street in Brooklyn. His father was teaching chemistry at Columbia University, but he spent many of his off-hours at the Fifty-Second Street Chess and Checkers Club in midtown Manhattan, a dark and mysterious cavern packed with chess addicts over whose bent heads hung suffocating levels of cigar, cigarette, and pipe smoke in several discrete atmospheric strata. Joe Weininger often competed there as an amateur, sometimes playing against the masters or even the grand masters of chess.

In 1954 the family moved to Schenectady, New York, where Joe Weininger had taken a job in the research division of the General Electric Company. His son Dave had acquired his own inclination for science and technology by the time he was a teenager, and when he was in high school he enrolled in a GE-sponsored electronics program. One of his early projects involved fixing a pulsed radar set that had been scavenged from an old B-19 bomber. When he was finished with the unit, it worked like new.

Dave's true love was chemistry, however, a taste he'd picked up from his father, who used to tutor neighborhood high-school kids at the Weininger home after work. Dave would sit in on these sessions, asking questions, sometimes late into the night, and soon he had acquired an easy familiarity with chemical compounds, formulas, and vocabulary. He had the chemist's natural ability to look at the formula $C_{12}H_{22}O_{11}$ and know immediately that this was sucrose (i.e., table sugar), and to look at the formula C_2H_6O and realize in an instant that this was ethanol, or as it was known in the outside world, alcohol, "the world's favorite drug," as he came to regard it. Dave absolutely loved the way in which the short and finite list of chemical elements in the periodic table was rich enough to have created all the items in the physical universe. Just a handful of chemical elements had sufficed to create everything that existed! At the same time, he realized that most others did not share his enthusiasm over this fact.

"It always surprised me afterward to find that most people in the world don't love chemistry," he said much later. "It's not even on their '100-most-beautiful-things-in-the-world' list."

Even at the tender age of about ten or twelve, it seemed to him that other people's failure to appreciate chemistry's true worth was essentially a language problem. Chemistry was one of Dave's native languages, a highly unusual talent not enjoyed by many of his peers. "My friends didn't love the way atoms went together to form paradimethlylaminobenzaldehyde—and with a name like that, who could blame them?" he said. "So from the age of about twelve or fourteen, I used to fantasize about a perfect chemical language."

He would invent just such a beast several years later. Meanwhile, with the help of his father, Dave started translating from German the autobiography of his boyhood hero, who, as it happened, was not a baseball player, football player, basketball star, or any of the other customary idols but instead was an organic chemist by the name of Emil Hermann Fischer.

Fischer had been one of the true creative geniuses of the discipline, although, unlike Dave, he was a late bloomer in chemistry. Fischer was the son of a merchant, and his father had wanted the boy to enter the family lumber business. Emil even tried it for a while, but without success, whereupon his father kindly informed him that he was too stupid to be a businessman and he would be far better off as a student. Soon Emil was at the University of Bonn attending the lectures of Friedrich Kekulé, the structural chemist who, according to legend, had a dream in which he saw the benzene molecule appear in the form of a snake biting its own tail. It was a prophetic vision, for the molecule did in the end prove to be a hexagonally shaped ring compound.

Later, Fischer would make several important discoveries of his own. He undertook a systematic study of the sugars (sucrose, lactose, maltose, and so on) and found that the best-known types contained six carbon atoms and could exist in sixteen different varieties, depending on the specific arrangement of their constituent atoms. Then in a series of experiments well known to chemists, he discovered that each separate molecular arrangement bent the rays of polarized light in a distinctive manner, and showed that it was the three-dimensional structure of the molecules that determined the way the light bounced off them. It was in this manner that he discovered the chiral nature of sugars, the fact that they existed in two main forms, a right-hand series and a left-hand series, each member of the pair being identical to the other except that each was the exact mirror image of its counterpart.

Fischer was also a master at synthesizing natural substances from their chemical components, and he synthesized caffeine, the stimulant in coffee and tea, and theobromine, the stimulant in chocolate, by combining the proper chemical ingredients. He even synthesized the proteins, the very building blocks of life. Chemically, any given protein was a combination of amino acids, and Fischer pioneered reliable methods of separating and identifying them, discovered two new amino acids in the process (proline and oxyproline), and performed several other minor miracles of protein synthesis. For his various syntheses and discoveries, Fischer won the Nobel Prize for chemistry in 1902, only the second such Nobel ever awarded.

By the time he got through translating Fischer's life story, Dave Weininger seemed to be fairly well launched into the career of an organic chemist. He looked fondly on the prospect of working in the GE industrial chemistry laboratory alongside his father and following in his footsteps. At that point, however, another side of reality reared its head in the form of literature, war, and a succession of somewhat lurid journalistic tales of the American "military-industrial complex."

Dave had started reading Kurt Vonnegut's novels—*Cat's Cradle* and *Player Piano* among them—dark, apocalyptic, dystopian books that gave him a whole new perspective about the appeal, or lack of it, of working for a faceless corporation while living in suburban Schenectady with the wife and kids. Soon he began to regard this as an embarrassingly limiting life plan. Being a chemist for GE, he now imagined, would be like living on a gigantic corporate plantation, a farm for the production of industrial chemicals, with he and his father as but two of the slaves.

Then finally in the late summer of 1968, just before he turned sixteen, Tom Wicker, the *New York Times* columnist, gave a lecture at Union College, which was not far from where the Weiningers lived. Dave was a big fan of Tom Wicker's; he read his columns every day they appeared, which was three times a week. So Dave and his younger brother, Art, bicycled down to the college to hear the talk, which was about the war in Vietnam.

For young Dave, this was a shocking, eye-opening, life-altering experience. Suddenly, the scales fell from his eyes, and he summarily concluded that if he weren't careful, he himself would wind up as a cog—or cannon fodder, or worse—in the great American war machine.

Bicycling back home in the dark, Dave Weininger made a series of critical life decisions. He decided to forget about going to college, becoming an industrial chemist, and working for GE. In fact, he decided to quit high school altogether, and most portentously of all, he even vowed to leave home. A week or so later he put together a few choice possessions, got himself a Suzuki R6 motorcycle (a model known as the "Hustler"), hopped on, and lit out for Colorado.

WHEREAS DAVE WEININGER, at least initially, wanted to learn one thing, chemistry, supremely well, Stuart Kauffman, his latter-day counterpart on the Info Mesa, had wanted from the very beginning to understand absolutely everything. And as if that weren't enough, he wanted to understand it deeply, totally, and completely, not just superficially, partially, or halfway—as Darwin, for example, had done.

Darwin had understood everything too, in a way, Kauffman thought, but he had left far too much unexplained, and even what he *had* explained, he'd done inadequately, having failed to go back all the way to first principles, down to the roots. Kauffman would go farther, perhaps even finish the job. He would penetrate more deeply into the nature of things; he would account for all of creation from the ground up, explaining the origin of life, the fundamental laws by which complex systems operated, the very existence of order in the universe. Where Darwin and all others had failed, Stu Kauffman would succeed.

Of course, Kauffman never voiced these humble aspirations exactly as such, even to himself, not in so many words. Rather, it was his practice as a scientist to start off by considering a certain narrowly defined and highly specific question—the problem of cellular differentiation of the zygote, for example, the way a newly fertilized cell developed into a differentiated organism with all kinds of specialized and structurally distinct parts. He'd then perform a short course of research on the topic, after which he'd go off to think it through on his own. But his mind was of such a nature that Kauffman tended to see any specific problem as only a special case of a more general scientific question. He'd see the problem of cellular differentiation, for example, as really, at bottom, the question of genetic regulation. And he'd see that in turn as the problem of how things organized themselves in general—not merely biological entities but nonbiological systems too, all kinds of things, everything. The problem of cellular differentiation, in other words, was actually the puzzle of self-organization itself, so that what had started out as a well-defined, small, and manageable scientific inquiry became in the end an all-embracing metaphysical quest.

Not that he found metaphysical quests daunting in the least. Stu Kauffman saw himself as always up to the challenge, whatever it was, and never seemed to be in over his head at any point. He wasn't drowning, awash, or even slightly at sea. He was on top of it all, even if often swimming against the tide.

It was also characteristic of his idiosyncratic approach to science that Kauffman often knew, in advance, how the answer would fall out—or at least how he *wanted* it to fall out. He tended to see the world through a particular aesthetic lens and harbored a preference for certain specific types of explanation as opposed to others. He had a distaste amounting to an abhorrence for the accidental, the fortuitous, the unlikely. He much preferred for things to be fated, inescapable, in the stars—especially the existence of life in the universe. If there was any one thing that Stu Kauffman didn't want to be accidental, it was the origin of life.

And as a final distinguishing characteristic of Kauffman's scientific modus operandi there was, as in Weininger's case, the matter of nomenclature, the preference for a distinctive use of language. For although he knew the King's English perfectly well, Kauffman did not make it an inflexible rule to always speak in the vernacular where science was concerned. This is not to say that he favored bafflegab or double-talk, although his terminology was often hard to understand at a first hearing. He simply made a habit of using the idiom that he felt was most appropriate to the problem at hand, and in view of his bent for abstractifying any given problem to the fullest, taking it at the highest possible level of generality, his language tended toward the abstruse, the obscure, and the metaphorical, almost as if he were speaking in code or, worse, parables. And so instead of talking about methods of finding solutions to certain types of problems, he'd talk about *strategies of simulated annealing*. Instead of talking about breaking a problem down into its parts, he'd talk about the *logic of patches*. Instead of talking about organisms filling biological niches in specific ways, he'd talk about *rugged and multipeaked models of fitness landscapes*.

He would use such terms and phrases not merely in print, in a technical paper, or even in formal lectures—which in Kauffman's case were always rather informal anyway—but even in face-to-face, let's-have-coffee, watercooler-type casual discussions, where he might blurt out something on the order of "Well, but you know in the NK model, each genotype is a vertex of a Boolean hypercube, adjacent to N other vertices, so what difference does it really make, you know what I mean?" and think nothing of it at all. Indeed, it would hardly be possible for a person to rise to a higher level of linguistic or conceptual abstraction while yet remaining even tenuously connected to the empirical realities of planet Earth.

From all of which one can gather that Stu Kauffman was deep, extremely deep, not your ordinary scientist, not your run-of-the-mill inquirer into nature. Whereas another researcher might make an entire career out of explaining the genesis of certain motility traits in slime molds, Kauffman might start out with some such little narrow and confining issue (in fact, he'd had an early publication called "A Mitotic Oscillator in the Slime Mold *Physarum polycephalum*"), but to him that was merely a point of departure, an opportunity to get into broader questions, square one for an enlarged, generalized, semimetaphysical disquisition in which much sooner rather than later, he found himself explaining not only the original problem but also order, lawfulness, everything!

But it had always been his practice to tackle large problems in his own way. His first major project while growing up in Sacramento had been to embark on a vast course of boatbuilding. He built model boats, full-size boats, canoes, barges, sailboats, power boats.

"They characterized me as a scientist," he said of these early efforts. "They were all grandiose and I refused to find out how to really make a boat by doing something sensible, like getting plans for the boat and then building it. I just made boats and they kept failing in various different hysterical ways."

He started doing this when he was just six or seven. His father, who had immigrated from Rumania in 1903, had been a real estate developer, among other things, and periodically brought home piles of scrap lumber. So Stu would take a few of these boards, knock them together until they looked about right, and then launch the resulting craft on the waters, whereupon it would sink without a trace. Seaworthy though they might have been to his untrained eye, each of his nautical creations concealed some tiny technical flaw—bad glue, weak wood, slightly askew jointures, whatever—that literally sank the final product on its maiden voyage, sometimes even with people aboard.

"What it shows is a predilection for having my own bright idea, not quite knowing the subject matter, and going out and trying it," he said.

Nor did he ever change his stripes. "If you find out what the experts think in a field, and read assiduously, and study very hard and are utterly responsible, then you will become trapped by what the experts tell you are the right questions, the right answers. You have picked up all of their presuppositions, studied them hard, locked them into your brain, and you're frozen by it."

Such an approach was not for him. Kauffman had never let himself get frozen into any one concept, theory, plan, or course of action. His earliest ambition was to be a playwright, and not just any ordinary playwright but "a *great* playwright," a Sophocles, a Eugene O'Neill, wrestling with profound thoughts while smoking endless pipefuls of Flying Dutchman tobacco.

He started writing plays while he was still in high school, beginning with a musical. "Aristotle defined two forms of drama, the tragedy and the comedy," Kauffman said. "But I added a third form, the atrocity."

He was not without a sense of humor. But what he was really interested in, he decided at length, was not so much the drama itself, not stagecraft, greasepaint, or the theater per se, but rather the ideas involved, the profundity and wisdom, the insights into human nature and the cosmos. So he went to Dartmouth, majored in philosophy, and graduated in 1960, summa cum laude, third in his class.

Then he went skiing. Kauffman was a great one for the outdoors—he loved hiking, mountain climbing, skiing, and mushroom collecting. So before accepting his Marshall Scholarship at Oxford, he took a vacation in the Alps, living out of a Volkswagen bus parked in the back lot of the Hotel Post in St. Anton, Austria. He spent his days grooming the slopes, climbing the peaks, and having a whale of a time.

Eight months later he rolled into Oxford, registered at Magdalen College, and got the shock of his life. "It was the first time in my life that I was surrounded by people who were smarter than I was," he said. Indeed, the place was crawling with Rhodes Scholars, people like David Souter, the future Supreme Court Justice, and George Will, the future columnist—not even to mention the professors, whom he found were almost supernaturally knowledgeable and accomplished.

Philosophy, however, was a great disappointment. You couldn't *prove* anything in philosophy, he decided. It was all theory, arguments, lots of people talking past each other down through the centuries and never getting anywhere in the process. Kauffman wanted to *understand* the world, to reach some fundamental conclusions about it, some *truths*, and for this he needed facts. Where better to get them from than in medical school?

"I figured somewhere I had to learn a bunch of facts, and if I went to medical school the bastards would make me learn a lot of facts," he said. "And that's exactly what happened."

It was in 1963, as a premed student at the University of California, Berkeley, when Kauffman first confronted the so-called problem of development, one of the chestnut puzzles in biology. The problem was how a single cell, a zygote, a fertilized ovum, developed into a differentiated organism instead of one big mass of homogeneous tissue. A fetus started out as a single cell, which then divided into two daughter cells, each of which divided into two others, and so on. If that kept on indefinitely, with each new cell being an exact carbon copy of its parent, the result would be an undifferentiated blob of meat. In actual fact, however, each of the cells somehow knew exactly what it was going to become, so that a predefined subset of cells developed into the brain, others into the lungs, still others into arms, legs, hands, and so on. How did the embryonic cells know enough to do that? It was a problem as old as Aristotle, and it appealed to Stu Kauffman as one of the great and primal mysteries. And so of course he had to tackle it himself.

At that point, in the mid-1960s, two French biologists, François Jacob and Jacques Monod, were putting forward some findings that suggested a solution. The key to it all lay in the genes inside the embryonic cells.

All of a cell's activities, including self-replication, were controlled by its genes. In the developing embryo, each cell possessed the same set of genes, but different cells developed into different bodily parts because distinct gene sequences became "active" in the various individual cell types. The question was, what made some gene sequences become active while others lay dormant? Jacob and Monod's answer was the genes themselves.

Certain molecules within the cells, they discovered, had the capacity to switch on certain genes while keeping the activity of others suppressed. The switched-on gene sequences would direct the cells to produce whatever the sequence coded for—hemoglobin, for example, or muscle fibers or nerve cells—while the rest of the gene essentially bided its time.

The most remarkable aspect of the whole arrangement, however, was that the controlling regulatory molecules did not originate from outside the cells but were themselves produced by the genes. In other words, the genes produced the very molecules that switched other genes on and off. The genes were, in a sense, controlling themselves, but only indirectly: gene 1 produced a molecule that switched gene 2 on or off, and vice versa. Collectively though, the genes controlled their own activities, and as a consequence, they regulated the course of cellular development.

To Stu Kauffman this was an extremely interesting result: there was evidently a little self-help, self-controlling, bootstrap mechanism at work inside the cells. Immediately, as was his wont, he broadened this narrow finding, enlarged it, and made it more abstract. This little bit of self-controlling genetic circuitry, he decided, was but a specific instance of a more general type of phenomenon, a special case of the way in which *any* given network of on and off states organized itself and gave rise to larger patterns of behavior.

What were the rules here? he wondered. How did these self-organizing systems operate? How big did a given network have to be before you got some novel behavior out of it?

These were not things they taught you in med school, or, so far as Kauffman knew, anywhere else, and so he decided to work out the problem on his own. He imagined the genes and their activities as light bulbs, and their on or off states as their being lit or unlit. He then said that a given light bulb A could switch light bulb B on or off, and vice versa. Finally he asked the question, if you put a bunch of these light bulbs together in a circuit, what would happen?

He tried to work it out on paper. He drew diagrams showing the light bulbs, patterns of crisscrossing wires, on and off states, and so on. These quickly got out of hand, so instead he made little tables that systematically charted all the various possibilities. Suppose, for example, you had three light bulbs, each connected to two others, with mutual switching activity taking place among them according to the rule that if the first two bulbs were lit at time t_0, then at time t_1 the third one would be lit also. At the next tick of the clock, t_2, all the bulbs would take the states of their nearest neighbors as new inputs and respond accordingly. As the clock marked the time from one moment to the next, a certain sequential pattern would emerge as the respective lights blinked on and off.

What Kauffman was investigating here was not anything so lowly and prosaic as mere light-bulb patterns or even genetic switching circuitry. No. What he was really looking into was the genesis of novelty itself, the birth of complexity, the rise of order.

Naturally, Stu Kauffman always expanded the system, considering ever bigger and more complicated light-bulb circuits. He produced his charts and tables incessantly, obsessively, even during classes, where, when he wasn't taking notes, he'd be sketching out blinking patterns and regulatory sequences in the margins of his notebook. He was by this time in med school, at the University of California, San Francisco (UCSF), and his grades were suffering accordingly: he got a C in pharmacology, a D in surgery. But it didn't matter! Who cared? There were more important things in the world, Horatio, than people's miserable foot problems!

He finally decided that he wanted to know what would happen if you had a system composed of *one hundred light bulbs*, each of which could switch some or all of the others on or off. Even Stu Kauffman, expert as he had become at all this, couldn't draw the necessary charts. For this he needed a computer.

Across the street from the med school there was a computer center where you could purchase data-processing time by the minute. This was now 1965, the early-Pleistocene era of the computer industry, where the IBM card still reigned supreme. So Kauffman learned how to program his light-bulb problem on IBM cards. He punched the cards so that they contained the relevant data, piled them into a neat stack, went over to the computer center, and handed his pile of cards to the clerk.

IN 1968, WHEN Stu Kauffman was getting his medical degree from UCSF (and graduating near the bottom of his class), another physician by the name of Anthony Rippo (who would later become the head of Bioreason, the Info Mesa's chief pharmacological data-mining firm) was just starting his latest new company. Rippo, it turned out, made a specialty of starting companies; in fact, he was irrepressible at it—he could hardly refrain from the activity even if he wanted to.

Rippo, a trim and smallish man with Ben Franklin–style rimless glasses, had started his first informal businesses while still in high school, "to get a revenue stream," as he put it, to support on-campus activities. He made ribbons to sell at football games, he threw ice-skating parties, and he made and sold assorted other products and services to generate a positive cash flow, something he's been doing ever since.

Both his father and grandfather were fishermen, which was to say that they too were entrepreneurs, small businessmen. From the time he was thirteen, Rippo himself went to sea on his father's tuna-fishing boat, the *Southern Queen*, which was docked in San Diego. He had even given some thought to making tuna fishing his career, and for a while seemed to be doing so, but for the sake of placating his mother, who had higher aspirations for her son, he'd finally enrolled in medical college. He got his medical degree from Loyola University in Chicago in 1966, came back to California, did an internship at UCLA, and then spent two of the Vietnam War years in the Navy, stationed at Long Beach.

Being a doctor didn't stop him from going to sea in ships, however, and he still went out on his father's tuna-fishing boat whenever he got the chance. Every so often, while hauling in tuna, miles at sea, Rippo would get radio calls from other boat captains asking for help with a sick crew member. They'd report the crew member's signs and symptoms, and Rippo would listen carefully, ask questions, and often be able to diagnose their ailments, sight unseen. But he realized how much better it would be to have a visual picture of the problem in front of him, and so with his developing entrepreneurial talent (by this time he and his father had already started a company that built and sold apartment houses), Anthony Rippo got the idea for a grand new business. He'd equip ships at sea with television cameras that would send back live pictures of sick seamen to Rippo's office. After looking at the images, perhaps requesting a close-up shot of this or that bodily part, Rippo would make an informed differential diagnosis of what the victim was suffering from. He could then suggest appropriate treatments, even to the point of prescribing drugs from the supplies that shipowners normally kept aboard their vessels for use in emergencies. Since ship captains often had no inkling of exactly what medicines were in their stockpiles, Rippo could also ask to see close-ups of the labels on their various vials, boxes, bottles, and tubes.

He called this latest company Marine Medical Services, a firm that proved to be a pioneer of the new field called *telemedicine*. There were a few snags in getting the operation off the ground—he had to get a waiver from the Federal Communications Commission to transmit data over voice lines, for one thing—but the red tape involved seemed to be no worse than that for starting any other business. Soon he had a contract with an ocean mining ship called *Deep Sea Miner II*, and by 1975 he was receiving slow-scan television pictures from it and other boats in the comfort and warmth of his medical offices. All this was quite new at the time, and Rippo considered himself altogether forward-looking and ahead of the curve.

Unfortunately, all this business activity had taken its toll, and in 1972 he and his first wife divorced. He remarried two years later, however, and he and his new spouse, Madeline D'Atri, both of them being good Catholics, would wind up raising six kids.

For all his love of starting companies, Rippo had never received any professional training in the art until much later, in 1981, when he moved to Rancho Santa Fe in northern San Diego County and met a venture capitalist by the name of Bill Norgren. Norgren was a master at starting successful businesses: having funded something like thirty to forty companies in his day, he put Rippo to shame, but he was happy enough to show Rippo the ropes.

At that time, venture-capital firms were hot to make medical clinic deals, for it seemed there were piles of money to be made in health care. So Rippo wrote up a business plan for a firm concentrating on industrial health in port cities. This enterprise got funded too and quickly became a thriving business in San Pedro, the main commercial port of Los Angeles.

Rippo lived in an apartment above the examining rooms of the new firm, which was called the Anderson Medical Clinic, and managed to get by on the cash flow from the company. Soon the clinic was such a success that it had branch offices in Houston and New Orleans.

After a while Rippo sold that company at a reasonable profit, and then at the behest of the Standard Club, an insurance company in London (known there as a protection and indemnity club), he started a medical service for merchant seamen around the world. The Standard Club would write policies for merchant seamen so long as Rippo would provide medical services for them while at sea. This turned into another successful venture, Marine Medical Management, and led to a substantial reduction of injuries and medical claims, allowing premiums to be cut in half within two years.

Anthony Rippo sold that company too, at which point he abandoned the formal practice of medicine. His strong suit, it had become clear at last, was not in healing people but rather in starting companies. Why not just start them, sell them, and be done with it?

WHEN STU KAUFFMAN dropped off his IBM cards at the computer center, he had scandalized the clerk by shuffling the stack, as if he were about to play a round of whist. "That's no way to treat your program cards," the clerk told him. "They have to be in perfect order."

But Kauffman hadn't shuffled the *program* cards, he'd only shuffled his *data* cards, those that represented the on/off states of his one-hundred-light-bulb switching network. He wanted to be sure that *those* cards were

in random order, or at least as random as he could get them. The program
cards were in their own separate stack, arranged precisely.

He didn't know what to expect from this computer run; in fact, nobody
did: so far as Stu Kauffman knew, the problem had never been run before.
Conceivably, a network consisting of a hundred light bulbs wired together
could wander all over the place aimlessly, at random, forever. Or maybe it
would freeze up immediately and stop dead—who knew?

What actually happened when he got the results back was that the light-
bulb system had almost immediately settled down into a characteristic
cycle of on and off states: many of the bulbs remained off most of the time
while many others remained lit. But a few of the bulbs—about eight or ten
of them—fell into a cyclical pattern of recurrent blinking that repeated
itself again and again, regular as the sunrise. The pattern made no inherent sense—it didn't look like anything recognizable—but it was obviously
a discrete and repeated motif and not just random noise. It was as if the
lights on a movie marquee were spelling out the same unknown coded
message again and again, although they had been programmed to do nothing more than flash randomly. Somehow, by means that were not clear, the
light-bulb network had acquired a life of its own.

The experience was one of the major historic moments of Stu Kauffman's career as a scientist, a result that has impressed him ever since. "I'm
still deeply proud of that," he said recently. "I'm still stunned that if you
make a random network with light bulbs and everybody has two inputs per
light bulb, and otherwise you make everything at random, the thing
behaves with order. Still blows me away! Thirty-seven years later, still
blows me away that that's true!"

Back then, he imagined that he was seeing something new and
extremely important, that he had in fact discovered some sort of novel
operating principle at the base of nature, but as yet he didn't know what
it was, how it worked, or anything else. He was sure, however, that this
was not just some quirky isolated result but something truly significant.
Something deep. (Stu's light-bulb networks were in fact an early, primitive version of physicist Stephen Wolfram's cellular automata, systems
that became a big hit in the spring of 2002 when Wolfram's self-
published book, *A New Kind of Science*, appeared on the scene. But
whereas Wolfram had acquired cellular automata from their inventor,
John von Neumann, Kauffman was discovering these similar, albeit

more elementary, structures, networks, and behaviors on his own.)

The important thing about Kauffman's experiment was that he had conceived of his hundred-light-bulb network as an analog to the way genes
worked; he had conceived of it, in other words, as a *simulation*. His original motive had been to find out what sort of order, if any, you might get
from a bunch of developing cells controlled by a hundred randomly acting
genes. If his computer results could be believed, and if they could be
extended to the actual operation of biological genes, then it would follow
that a batch of a hundred genes, even without the benefit of any specific
program having been encoded into their chromosomes, would give rise to
some kind of order, automatically, spontaneously, "for free."

Well! This was certainly more fascinating than mere medicine!

Additional computer runs with bigger networks, with as many as four or
five hundred light-bulb "genes," convinced Kauffman that a power law
was at work, for he soon discovered that the number of light bulbs
involved in the orderly flashing patterns varied as the square root of the
total number of bulbs in the system: one hundred light bulbs, for example, gave rise to a ten-bulb flashing sequence; a five-hundred-bulb system
gave you twenty-two or so blinking bulbs, and so on.

Lo and behold, he then learned that a comparable power law was at
work in biology as well, for his textbooks told him that the number of cell
types in an organism scaled roughly as the square root of the number of
its genes. The human body, for example, was thought at that time to be a
product of some 100,000 genes (later estimates reduced the figure to
30,000 genes), and the body was known to be composed of something like
256 different cell types. The fact that approximately the same power law
held true in the two cases, Kauffman thought, was probably not sheer
coincidence. There were some passably deep meanings in all this, he
decided.

With his discovery of a hidden source of order in genetic switching networks, perhaps he'd even supplied something that had been missing from
Darwinism when considered as a complete account of the order to be
found in nature. Kauffman was no anti-Darwinist—he was not a creation-
ist of any stripe, nothing like that whatsoever. Natural selection, he was
solidly convinced, had operated throughout the entire evolutionary history
of organisms—that much was clear from the fossil record and other evidence. Still, natural selection could not be the whole story, as even Darwin

himself had acknowledged, saying in the introduction to *The Origin of Species* that "Natural Selection has been the most important, but not the exclusive, means of modification."

Natural selection couldn't be the sole explanation of life on earth, Kauffman decided, the reason being that there hadn't been enough time since the beginning of the universe for selection, acting alone, to have canvassed, discarded, and preferentially selected for all the winning combinations from among the total mathematical space of possibilities. If natural selection were the only mechanism providing for order in nature, then far too much order had accumulated in the world than was theoretically possible in the time available. Therefore, something else must have been at work on a lower level—at the molecular level perhaps—a principle of order that had given selection a head start by providing it with some fine-tooled prefabricated source materials to work with.

Kauffman was thinking all these deep thoughts entirely on his own while still a medical student and taking all the required courses in anatomy, oncology, comparative morphology, and everything else. By now, however, medicine had taken such a clear backseat to Kauffman's own private universe of interests, his research into the origins of order, that it was fortunate in the extreme that the UCSF medical school allowed their students to take three months' leave in order to pursue a separate study program elsewhere. In 1966 Kauffman chose the Massachusetts Institute of Technology (MIT) and Warren McCulloch.

Warren McCulloch was one of the founding fathers of the information age, a theorist who was in large part responsible for the view that mental activities were a form of information processing. He, like Kauffman, was absolutely crazed about *networks* and seemed to consider them almost more real than the objects of which the networks were composed.

McCulloch himself had something of a checkered past, having acquired a medical degree from Columbia in 1927, after which he taught psychology, worked at Bellevue Hospital in New York City (an institution for the mentally ill), and then wrote a book called *The Isocortex of the Chimpanzee*. Later he became a staff member at the Research Laboratory of Electronics at MIT.

McCulloch's major claim to fame, however, was a paper he'd written with Walter Pitts, a student at the University of Illinois, where McCulloch had been director of the Laboratory for Basic Research in the psychiatry department. The paper was called "A Logical Calculus of Ideas Immanent in Nervous Activity" and was published in a volume of the *Bulletin of Mathematical Biophysics* in 1943. McCulloch was forty-four at the time; Pitts, a mathematical prodigy, was eighteen. Between them they argued that human mental activity was formally identical with an abstract network of functions such as *and*, *or*, and *not*, and could be simulated by a suitably programmed or wired machine. In other words, anything a human being could do by manipulating symbols according to a codified set of rules could be done equally well by a totally mindless mechanical appliance.

They argued by analogy. The human brain was a network of neurons, each of which was connected to a large number of others—one cell wired up to as many as 200,000 other neurons. Each neuron was in an on or off state, either active, firing electrical impulses down an axon, or inactive, not firing them. Moreover, a given neuron's activity was controlled by the inputs it received from the others to which it was connected. If the impulses it received were greater than a certain threshold value, it would fire; if not, not.

Pitts and McCulloch then showed how the identical behavior could be reproduced in an idealized network made up of artificial black-box "neurons," whose activities were controlled by a fixed number of others. The brain, in a sense, was itself a sort of machine.

This eventually became the orthodox viewpoint among artificial intelligence researchers, but it was all the more impressive at the time because modern electronic digital computers did not exist when Pitts and McCulloch were advancing their radical if not heretical thesis about the operation of the human brain. Anyway, it was McCulloch's work with neural networks that made Stu Kauffman want to work with him.

Kauffman wrote a letter to McCulloch in Cambridge telling him about his genetic simulation results and asking if he could come up to MIT to study with him for a couple of months. "All Cambridge excited about your work," McCulloch wrote back. Modesty not being Stu Kauffman's middle name, he took McCulloch's comment literally; later on he understood that this was just an example of the master's often rather baroque manner of speaking.

Kauffman arrived in Cambridge with Elizabeth Ann Bianchi, an art historian whom he'd met while at Oxford and had just married. The two of them ended up staying with the McCullochs.

Stu spent his days at MIT simulating ever larger genetic networks on the Project MAC computer (Multiple-Access Computer, an early time-sharing system) and talking with McCulloch and with Seymour Papert, Marvin Minsky, and the rest of MIT's artificial intelligentsia, finally coming away with the view that his own genetic networks and McCulloch's neural nets were at base one and the same.

In 1967, Kauffman and McCulloch wrote a paper together, "Random Nets of Formal Genes," the first scientific publication on the subject, which was issued as a quarterly progress report of the MIT Research Laboratory of Electronics.

"Warren, is anybody going to care about all this?" Stu asked at the time.

"Yes, but it will be twenty years before anyone notices it," McCulloch answered—which proved to be about correct.

3

THE EXTINGUISHED PROFESSOR

ON THE MORNING of November 1, 1952, George Cowan stood on the gently rolling deck of the USS *Estes*, looked toward the island of Elugelab thirty miles over the horizon to the north, and put on a pair of high-density goggles. At 7:15 A.M., the device inside the shot cab changed from an inert collection of steel containers, wires, and dead chemicals into something resembling a newborn star. The detonators of the first fission bomb (there were two of them inside the device) fired simultaneously, imploding the uranium shell to a critical mass, setting off the initial fission explosion. The shock wave from it heated and crushed the liquid deuterium inside the Sausage and set off a second fission bomb known as the *spark plug*. This second blast in turn ignited the thermonuclear reaction itself. The hot plasma burst through and then vaporized the steel casing of the device, and a fireball more than three miles in diameter appeared on the horizon.

Cowan felt a wave of heat and light on his face, as if someone had opened the door to a furnace.

"I was stunned. I mean, it was big," he said much later. "I'd been trying to visualize what it was going to be like, and I'd worked out a way to calibrate the shot. As soon as I dared, I whipped off my dark glasses and the thing was enormous, bigger than I ever imagined it would be. It looked as though it blotted out the whole horizon, and I was standing on the deck of the *Estes*, thirty miles away."

The bomb had erupted with a force of 10.4 megatons, equivalent to roughly a thousand Hiroshimas. The fireball ripped through the helium box, rapidly consuming all nine thousand feet of it. The mushroom cloud rose more than 100,000 feet into the air, expanded to a final diameter of more than one hundred miles, and rained mud, water, and radioactive debris back down on the atoll. The island of Elugelab was summarily removed from the face of the earth. Where Elugelab had been, there was now only a crater a mile wide and two hundred feet deep.

Prior to the test shot, Cowan had placed two sets of fast-neutron detectors on the Sausage at Elugelab. He'd installed one inside the casing, and he'd affixed the other, a backup, to the outer surface of the device. In the immediate aftermath of the explosion Cowan saw that both detectors had telemetered their readings back to the *Estes*, but at a first glance the numbers did not agree. He would have to resolve the discrepancy later, for a helicopter was about to leave the deck of the *Estes*, and he was scheduled to be aboard it.

The helicopter took him to a nearby aircraft carrier, where he climbed aboard a dive bomber. Soon the dive bomber was shot off the deck by steam catapult—Cowan's second amazing experience of the day.

An hour or so later the plane landed at Kwajalein, about four hundred miles east of Elugelab. Two F-84 Thunderjets that had flown through the edge of the mushroom cloud at forty thousand feet to take radioactivity measurements and to collect samples of airborne debris had landed there just before him, bringing their hot cargo with them. The samples were put inside heavy shielded containers, which were then loaded aboard a C-54 cargo plane headed for New Mexico. Soon Cowan was on that plane too, still dressed for the tropics.

When the C-54 finally landed at Albuquerque, there was snow on the ground and more of it was falling. With no other clothes, Cowan wrapped himself in an Air Force blanket and staggered out onto the tarmac, the first person from Eniwetok Atoll to make it back to the states.

An army truck stood by in the gloom, waiting for the sample containers. The driver loaded them into the truck and Cowan climbed into the front seat. He was sound asleep by the time they rolled into Los Alamos.

It took George Cowan a few months to figure out why he'd gotten two different readings from his neutron detectors. The internal detectors had given lower counts, he decided, because some of the neutrons produced by the blast had been captured by the nuclei that had been created during the course of the thermonuclear reaction. The detectors placed outside the device, not subject to such capture, had given higher and probably more accurate neutron counts.

The H-bomb test at Elugelab had abundantly confirmed the explosion simulations that had been performed by hand calculations and by the ENIAC, MANIAC, Institute for Advanced Study, and SEAC computers. Later analyses would show that events inside the device had taken place pretty much as the computer models had predicted they would. Well before the time of the Ivy Mike test shot, computers had already become permanent fixtures at Los Alamos.

Cowan stayed on at the lab for the next thirty-six years and for much of the time was part of the effort to make thermonuclear weapons ever smaller, lighter, and more reliable. It was the Cold War era, after all, and H-bombs, precisely because of their overwhelming destructive potential, could help to preserve the balance of power, hold other nations in check, and constitute a force for peace. Such was the thinking at the Los Alamos lab anyway.

Building, simulating, and testing bombs were not the lab's only mission, however: another one, ironically, was research into human genetics. That work had begun immediately after the Hiroshima and Nagasaki bombings, in an attempt to determine the effects of radiation on the genetic endowment of the survivors. A few of the world's more excitable scientists had claimed that the genetic mutations induced by atomic radiation could conceivably lead to a fundamental alteration of the race and perhaps its eventual destruction. In 1947, for example, the Caltech geneticist Alfred Sturtevant, writing in *Science*, predicted that the atomic bombs already exploded "will ultimately result in the production of numerous defective individuals—if the human species itself survives for many generations." At the time, such claims had to be taken at least semiseriously, and who better to investigate them than the Los Alamos lab's own scientists?

Later, in the 1950s, Cowan himself became interested in genetics, and molecular biology in particular, after he'd heard a talk on the subject by George Gamow, who had argued, just ten months after Watson and Crick had identified DNA's structure as a double helix, that the information stored in the DNA molecule was probably expressed in the form of a four-letter digital code. In 1978 Cowan became director of research at Los Alamos and presided over the gradual introduction of molecular biology into the lab's overall agenda. By the time he departed as head of research

in 1981, the Los Alamos National Laboratory would be participating heavily in preparations for the Human Genome Project, the attempt to
uncover the full set of DNA sequences found in human cells.

Over time, a new view of natural phenomena had begun to take shape
at Los Alamos. First under the influence of nuclear-bomb shock-wave
simulations, and then as a consequence of efforts to understand the
human genome, the lab's scientists subtly and imperceptibly shifted from
a linear to a nonlinear view of how the world worked, or at least some of
the more interesting items within it.

Linear phenomena were governed by the laws of classical physics, those
in which a given variable was a simple and easily predictable function of
another. For example, if you applied a given amount of force to a moveable object, the object moved away at a given speed, and if you increased
the force, you increased the speed by a proportionate amount. A graph of
the relationship would be a straight line, and for that reason such a function was said to be "linear." The equations of Newtonian mechanics
described all sorts of linear phenomena perfectly well: artillery trajectories, the paths of billiard balls, the flight of spacecraft, and the motions of
planets, stars, and galaxies.

Nonlinear phenomena, by contrast, presented a quite different face to
the inquirer into nature, for there were many natural processes, such as
fluid flow, the development of weather systems, the course of evolution,
the explosion of a thermonuclear bomb, the flowering of a plant, and the
development of an organism from a single cell to a finished animal, in
which the final result did not seem to follow in any simple, straight-line,
linear fashion from any one causal factor or, in fact, from any readily cal-
culable combination of them. Such phenomena, instead, appeared to arise
from the simultaneous interaction of many separate but converging lines
of development. But there were so many mutually intersecting variables
at work in such cases, all of them contributing their own separate share to
the final result, that no single equation could ever hope to cover such
things adequately; the only way to understand them would be by computer simulation of what was likely to happen when all the causal factors
came together, each of them kicking in with its own respective addition.

The fall of an apple was a linear phenomenon; the fall of an empire was
nonlinear. To George Cowan, it seemed increasingly as if more of what
actually went on in nature was nonlinear, the province of multidimen-

sional and mutually intersecting lines of causality, and that the linear was in fact an idealized academic construct that did not do justice to the complexity found in the real world.

By about 1981, Cowan had the idea for starting a new type of research institute, one that would address the actual variability and manifold nature of the world as found, as opposed to the world that hitherto had been studied by classical, billiard-ball physics. Such an institute would have to attract those who weren't hemmed in by the classical preconceptions, those who were willing to think outside the square. Since it would have to be housed in a decent physical facility and not operate out of a garage, it would also have to attract a reasonably large amount of seed money.

Cowan knew that lots of people had had similarly grandiose ideas in the past but that few of them had succeeded in getting beyond the inspiration stage. So he thought it would be wise to place this new scheme of his before some of his colleagues at Los Alamos.

AT LOS ALAMOS there was a tradition of weekly lunches of the so-called Senior Fellows, a gathering of the retired, the semiretired, and the decidedly over-the-hill. All of these fine scientists would meet in the lab cafeteria to talk over matters of surpassing global importance, such as their golf scores, for example. Nick Metropolis, the computer expert, the man who had built, named, and run the MANIAC computer and who had done the H-bomb simulations on that machine, was one of this select group, as was Pete Carruthers, Stirling Colgate, and some others. And of course there was also George Cowan himself.

Two years after he'd retired as head of the research division, Cowan broached to the assembled lunchtime fellows the idea of a new transdisciplinary, paradigm-shifting research center and discovered that his bold notion was an immediate hit with the audience.

Not that this was entirely surprising. The Los Alamos scientists, after all, were nothing if not au courant with the latest science and new technology, and among them the terms *complexity*, *nonlinear*, *chaos*, and their kin were even a bit old hat. All those present understood, for example, that *chaos*, the latest buzzword, was not to be taken at face value. The term did not refer to events that were chaotic in the sense of being ungoverned by the laws of nature: *nothing* was ungoverned by the laws of nature. Rather, the term referred to outcomes that were hard to predict because the course of events depended heavily on initial conditions and subsequent accidental influences. Smoke eddying up from an outdoor barbecue, the dynamics of lava flows, water-droplet patterns from a dripping faucet, the distribution of species within ecosystems—all these so-called chaotic phenomena were as firmly embraced by natural law as anything else in the material universe; they were just hard to predict in any detail. The members of the lunchtime group were in favor of studying such phenomena systematically precisely because those types of systems were omnipresent in nature and because there was as yet no academic institution devoted to investigating such events and processes. The Los Alamos fellows decided that there should be.

In fact they went further, proposing the notion that superficially similar phenomena lying outside the hard sciences ought to be brought under the same umbrella. The behavior of consumers in a market economy, the bunching and unbunching of cars in a traffic jam, the progress of an arms race, the rise and fall of civilizations—these too were examples of complex, nonlinear, or chaotic phenomena and were therefore fit subjects for Cowan's proposed new institute. Hard scientists could mix in with those of the softer variety—economists, political scientists, anthropologists, and the like—and one and all could engage in a heady mix of theoretical speculation, cross-fertilization, and collective ferment.

But there was a third necessary element, and that was large-scale use of computers, for how could such inherently complicated, multifarious, and otherwise intractable systems be studied without the aid of lavish and high-powered computer simulations? Computers by now had become such commonplace tools in science that it was as if the machines had been made expressly for scientific research, as indeed, many of them had been: the Institute for Advanced Study computer, the MANIAC, the JOHNNIAC (named for John von Neumann)—all of them had been intended for and used by the working scientist. Nick Metropolis and Gian-Carlo Rota, an MIT mathematician who was a visiting Los Alamos fellow, therefore saw the new institute as primarily a center for advanced computation. The others wanted the place to be something more than that but found it hard to specify exactly what, and even George Cowan himself never seemed to describe the proposed institute and its mission exactly the same way twice.

The other question was where to put it. The two most obvious places were Los Alamos and Santa Fe, the state capital. Santa Fe, as everyone knew, was a center of culture. It was an artistic community par excellence, overflowing with art galleries and museums, jewelry shops, bookstores, expensive hotels, fabulous restaurants, quaint adobe architecture, recitals, and the Santa Fe Opera, of which George Cowan happened to be cofounder and a board member. The place was full of tradition, lore, and history. On the other hand, there wasn't much if any science there to speak of. In fact, the city was fairly crawling with churches, convents, homes for unwed mothers, and parochial schools. That, together with the spiritual influence of the American Indian, plus an overabundance of "alternative" medicines, therapies, and approaches (there was a one in fifty-two chance that any given Santa Fe resident was a "healer" of some kind), gave the place a somewhat occult and otherworldly air.

Los Alamos, by contrast, was 100 percent hard science, period. Lab inmates were seemingly addicted to the sciences and had little apparent interest in anything else, at least if the surrounding community was any guide, for the city was largely made up of gas stations, motels, hardware stores, bowling alleys, private residences, and the airport. "Culture" hardly existed in any meaningful sense of the word.

Still, it was really no contest in the end. Los Alamos had been an artifact of national emergency, a chance collision of necessary secrecy and deliberate remoteness. No such constraints would apply to the freethinking, creative, outside-the-box institute that Cowan and his cadre of Los Alamos Senior Fellows had in mind.

Santa Fe was a paradise, pure and simple. It already boasted a couple of highly regarded academic institutions, the School of American Research, devoted to the study of Native American culture, and St. John's College, set on its own exalted mountainside overlooking the city. And so a vision of a grand new utopia of scientific meditation sprang up in the heads of the Senior Fellows: a bunch of latter-day Renaissance men bearing computers.

The final operational question was, where would the land and the money come from? Sums well into the nine figures—$100 million or more—were bandied about by the Los Alamos fellows, none of whom had access to any such vast fortunes. Cowan himself was not exactly impoverished (along with his other feats, Cowan also happened to be the founding director of the Los Alamos National Bank, the largest private bank in New Mexico), but he did not envision the institute as a monument to himself. And he had a few friends with some excess real estate holdings in the vicinity, but getting those people to part with their precious tracts of land for the sake of a nebulous new research institute was something else again.

Everything changed in the summer of 1983, however, when Murray Gell-Mann entered the picture. Gell-Mann was a legend, one of the last living icons of science, one of the haloed, a sort of nonbiological parallel to Watson and Crick. Trained as a particle physicist, he was regarded by some as a genius. As a youngster, he was precocious beyond belief. He entered Yale University in 1944 on the day he turned fifteen. He received his Ph.D. from MIT at the age of twenty-two. He was the author of the quark theory of matter (his Subaru station wagon bore the vanity license plate QUARKS), for which he won the Nobel Prize for physics in 1969, and he won it alone—it was not one of those multiple-person awards that later became the norm for the Nobel Prize committee. He seemed to know a million languages (his father had been a linguist) and was not above telling people the meaning and etymological derivation of their own names, or how to pronounce or even spell them correctly. "Anyone who knows me is aware of my intolerance of mistakes, as manifested for example in my ceaseless editing of French, Italian, and Spanish words on American restaurant menus," he once admitted in a rare burst of self-reflectiveness.

He was possessed of an ego of substantial dimensions and often spoke in witty put-downs and pithy wisecracks. Solid-state physics, which he detested, he referred to as "squalid-state physics."

He was a fearless punster. Assuming that there was a scientific principle beyond "general" relativity, he then had to wonder, "While we are waiting for it to be discovered, what should we call that principle? Field marshal relativity? Generalissimo relativity? Certainly it goes far beyond general relativity."

Sometimes the humor was even at his own expense, as when after retiring from Caltech, he took to referring to himself as an "extinguished professor."

Murray, as his friends called him, was beloved for it all. Marvin Minsky, clearly an admirer, once said of Murray Gell-Mann, "I think his major contribution is inventing new kinds of insults."

Anyway, when he heard from a mutual friend that George Cowan was thinking of starting an institute in New Mexico to foster the study of complexity theory, chaos, nonlinear dynamics, and the like, Gell-Mann could hardly believe his good luck. For one thing, as an elderly (if not yet defunct) particle physicist, he himself had developed an interest in such phenomena, which he saw as the next new thing, an "emerging synthesis" of hitherto unrelated fields. For another, he had just moved into his second dream house (his first had been in Aspen), located on a hilltop lot in Tesuque, north of Santa Fe. The prospect of a new-wave scientific research center springing up in his own backyard was almost too good to be true.

In 1983, he gave Cowan a call and offered what help he could.

Cowan realized at once that since Gell-Mann was a director of the MacArthur Foundation, a place that always seemed to be gushing fountains of money in every direction, his name, reputation, intellect, gift for gab, charisma, and connections could be a considerable amount of help to the new institute.

In the coming months Gell-Mann gave a few talks before Cowan's core group of institute supporters. These masterly performances gathered together the institute's themes in such a way that the others found irresistible. Nature was one unified whole, Gell-Mann said, whereas the extant scientific disciplines were specialized, fragmented, and tightly confined within their own narrow borders. Cowan's institute would blur the edges, merge the disciplines into a unique synthesis that was greater than the sum of its parts, and inaugurate a fundamental new approach to the understanding of natural phenomena. Computers, which of course were complex systems themselves, would simulate the various processes under investigation and reveal the common elements that lay hidden beneath the fabric of appearances. A deft union of the simple and the complex seemed to be in the offing.

Gell-Mann also made use of a portentous new jargon, as was only fitting, perhaps, to the birth of the wondrous new field he was describing. It was an ever grander mother tongue, filled with talk of complex adaptive systems, self-organization, self-similarity, sustainability, emergent behavior, and eventually, "plectics," Gell-Mann's own chosen show-off neologism for the emerging synthesis he had in mind. *Plectics*, in his view, would be a grand mix of everything from molecular biology to the big bang, along with a healthy dose of cognitive science, computer simulation, and save-the-Earth environmentalism thrown in for good measure (Gell-Mann was a great proponent of saving the condors).

As he himself had once tried to explain it, none too successfully, "What I like to say is that the subject consists of the study of simplicity, complexity of various kinds, and complex adaptive systems, with some consideration of nonadaptive systems as well. To describe the whole field I've coined the word 'plectics,' which comes from the Greek word meaning 'twisted' or 'braided.' The cognate Latin word, *plexus*, also meaning 'braided,' gives rise to 'complex,' originally 'braided together.' The related Latin verb *plicare*, meaning 'to fold,' is connected with *simplex*, originally 'once-folded,' which gives rise to 'simple.' "

Whatever it all meant, people were spellbound by the rhetoric, by the cerebral appeal of the man, by the dazzling intellectual panorama he laid out before them.

And so in March 1984, while the proposed institute didn't physically exist as yet and had no money, staff, land, or definite name, its group of core supporters appointed George Cowan as president of the place and Murray Gell-Mann as chairman of the board.

FOR THE SAKE of completing his physician's requirements, Stu Kauffman did a year's internship at Cincinnati General Hospital. There, for what it was worth, he practiced the traditional laying on of hands, delivering sixty babies, doing spinal taps on infants, tending to people in cardiac arrest—the entire gamut of modern curative witchcraft. Still, his body and soul weren't in it, this was not *his thing*, and so, just as he had earlier forsaken boatbuilding, playwriting, and philosophy, he now at once and forevermore abandoned the practice of medicine. In fact, a far more attractive opportunity lay ahead.

In 1967, after spending three months with Warren McCulloch at MIT, Kauffman had been invited to present his results on order, genes, and cell types at a scientific conference. Those experiences had been much more to his liking and would end up changing his life. The conference had been held at the Villa Serbelloni, a vast Elysian estate on Lake Como in Italy.

"It was just wonderful," Kauffman recalled later. "A site picked out by Pliny the Younger. Absolutely gorgeous."

Lake Como was a Y-shaped body of water, and the Villa Serbelloni was a group of pink buildings with stuccoed walls, red-tile roofs, and frescoed ceilings, the whole strategically placed on a hillside. The site offered commanding views of all three arms of the lake and of the mountains that trailed off into the misty distance. Just being there and beholding this vista made you feel like a Roman emperor.

"And here were all these amazing people!" Kauffman added.

One of them, Jack Cowan (no relation to George), ran a program in theoretical biology at the University of Chicago. Kauffman gave his talk, answered questions, and afterward he, Cowan, and some others went out and had coffee on the terrace overlooking the pool and beyond it, the blue waters of Lake Como.

The terrace was planted with flowers, shrubs, and tall trees. The lake stretched off toward the horizon. Kauffman was flush with excitement from the success of his talk, and at that transcendental moment, right then and there, Cowan asked Stu to join the faculty at Chicago.

Kauffman looked at Cowan. He looked out across the lake to the long green hills on the other side. He smiled.

"Of course!" he said.

Two years later, with his medical degree in hand and his residency behind him, Kauffman was in Chicago doing experiments on fruit fly genetics. Partly this was to get a firsthand view of the processes involved in embryonic development, to learn how living organisms grew and differentiated themselves and blossomed from zygote to finished animal. Partly it was to get some hands-on experience of a modern biology lab. But most fundamentally of all, Kauffman just wanted to soak up the available facts about how life worked. If he was going to be theorizing about the subject—he was at heart a theorist, he decided, but he always wanted his speculations to be solidly grounded in experimental reality— then it would be best to be acquainted with the facts of life from the ground up.

And so when he wasn't looking through the microscope at the latest fruit fly mutation, he let himself think again about the little genetic regulatory networks he had discovered. The reason those networks had so fascinated him to begin with was that they had showed how order could appear out of nowhere—"for free." You didn't need the hand of God to introduce the subtle patterning that cropped up; you didn't need a preexisting template; you didn't need any kind of gimmicky, zip-apart, zip-back-up, self-replicating molecule such as DNA. Rather, you simply started out with a bunch of primitive nodes and switching possibilities, you let the system run, and order arose of its own accord.

In the most general sense, such a network was an assemblage of simple entities whose on and off states depended on the similar states of other equally simple substances. You could hardly get more basic than that. But if you let such a system run, then order automatically appeared.

Without a doubt, this was a form of *creation*, Kauffman thought. But it was a creation from the bottom up, not one that had been imposed from the top down or from the outside. It seemed to him that with such a paradigm, he might well be able to explain the origin of life on earth.

Like the problem of development, the origin of life was one of the long-standing, chestnut problems of science, and Darwinian evolution offered no help in solving it. Natural selection, as Darwin intended it, was a system for the explanation of the *types* of living things on earth rather than an explanation of the *fact* that living things existed on earth. The theory presupposed life as a successful, preexisting, ongoing phenomenon.

However, the problem was that if life hadn't always existed, then it had to have arisen out of nonliving entities. But how could you get vibrant, pulsating life-forms out of inert clumps of dead chemicals? The very notion smacked of "spontaneous generation," a theory long since discredited by science. As Darwin himself had written in an 1863 letter to a friend, "It will be some time before we see 'slime, protoplasm, &c.' generating a new animal. But I have long regretted that I truckled to public opinion, and used the Pentateuchal term creation, by which I really meant 'appeared' by some wholly unknown process. It is mere rubbish, thinking at present of the origin of life; one might as well think of the origin of matter."

Darwin continued to think about the origin of life, however, and eight years after writing those words, he stated in another letter to another friend, "It is often said that all the conditions for the first production of a living organism are now present, which could ever have been present. But if (and oh! what a big if!) we could conceive in some warm little pond, with all sorts of ammonia and phosphoric salts, lights, heat, electricity, &c.,

present, that a protein compound was chemically formed ready to undergo still more complex changes," then, he implied, you would have produced the basic ingredients of living things.

Those were Darwin's preferred ingredients of life: a *warm little pond, ammonia, salts, light, heat, electricity*.

Some eighty years later, in 1952, two chemists at the University of Chicago had experimentally duplicated those very conditions. Harold Urey, the 1934 Nobel laureate in chemistry, and his graduate student, Stanley Miller, had constructed a closed chemical apparatus in which the "pond" was a heated flask of water, and supplies of ammonia, hydrogen, and methane—dead chemicals par excellence—were fed in over the water's surface. Steam from the heated water drove the chemicals around a loop that included a five-liter glass flask through which electrical sparks were made to jump between tungsten electrodes, simulating lightning. As well as they could imagine it, the chemists had reproduced the atmosphere and environment of the early, prebiotic earth.

Every so often, they drew off samples through stopcocks and analyzed them by paper chromatography. They did not find any "new animals" lurking in their artificial cosmos. What they did find were twenty new chemical compounds that hadn't been there at the beginning. Four of those compounds, moreover, were amino acids: glycine, alanine, glutamic acid, and aspartic acid, the building blocks of proteins.

The Urey-Miller experiment became a classic and was commonly cited ever afterward when scientists addressed themselves to the origin of life. As an explanation of life's origin, however, the Urey-Miller experiment suffered the fatal limitation that it had not actually produced *life*, only more dead chemicals (i.e., amino acids). Indeed, when Urey had been asked beforehand what he expected to make with his apparatus, he replied, *"Beilstein."* *Beilstein* was the name of a German reference text that listed all the organic molecules known to chemists. It was an encyclopedic work, then in twenty-eight volumes (many more were to come in the future), but still, it was a list of inert and lifeless molecules, not living things.

In Stu Kauffman's view, you needed far more than a bunch of newly created amino acids in order to explain the appearance of life on earth. Specifically, you needed a means by which the building blocks—the amino acids, the proteins, or other small molecular components—could at least reproduce themselves, could systematically create more of their own kind. And you needed a means by which those self-reproducing molecules could combine so as to give rise to a metabolism.

"How, without supervision, did all the building blocks come together at high enough concentrations in one place and at one time to get a metabolism going?" he wondered. There didn't seem to be enough time since day one of the universe for all that to have happened.

But there *would* have been, Kauffman decided, if the events had been speeded up somehow—by means of catalysts, for example. Catalysts, he knew, were molecules that accelerated the pace of chemical reactions without themselves being changed or consumed in the process. If the prebiotic soup had included a sufficient variety and density of catalysts, then all sorts of things might have been possible that wouldn't have been possible otherwise. In particular, catalysts might have speeded up the pace and complexity of chemical reactions to the point that self-reproducing molecular combinations, and even systems approaching the complexity of a primitive metabolism, were possible—and perhaps inevitable.

A metabolism, he thought, was a complex of self-sustaining chemical reactions. Such reactions were rare occurrences in chemistry. Two molecules might combine to form a third, or one molecule might break up to create two or more. The initial reaction products, in turn, might give rise to further molecular combinations, but typically those molecules did not spontaneously recombine so that the same reactions and molecular products kept recurring again and again, with the molecules in effect feeding on themselves as raw materials.

But now Kauffman came up with a clue as to how that might happen. *What if the molecules catalyzed themselves?* Not in the sense that a lone molecule A catalyzed itself (which was impossible), but rather in the sense that a network of molecules was large enough so that those of type A catalyzed the formation of others of type B, which in turn catalyzed still others of type C, and so on around a big loop until at the end, molecules of type Z ended up catalyzing those of type A. No given molecule *itself* would be self-catalyzing, but nevertheless the whole *set* of molecules would be.

This is how Stu Kauffman came up with his theory of self-catalyzing, or "autocatalytic" sets. If such things existed, then life would be a natural property of complex chemical systems. "When the number of different kinds of molecules in a chemical soup passes a certain threshold," he said, "a self-sustaining network of reactions—an autocatalytic metabolism—will suddenly appear."

To check this out, Kauffman now made some little pencil-and-paper models of how such autocatalytic sets could arise. He drew diagrams showing circles of various sizes (representing molecules), squares (which were chemical reactions), plus a web of lines and arrows denoting the direction of the reactions and pointing to the various molecular products. These models quickly became too complicated to follow on paper, so he simulated them on a computer. What he discovered was that once a system of molecules, catalysts, and chemical combinations reached a minimum threshold of size and complexity, then autocatalytic reactions indeed began to occur. Given enough chemicals, heat, and time, a primitive metabolism was almost certain to arise. Life was not only probable but also almost guaranteed.

That conclusion, added to his earlier results concerning the manner in which order arose spontaneously, led Stu Kauffman to what he considered to be his deepest realization yet about the way life worked. The world's basic constituents—molecules, catalysts, chemical reaction products—were not the inert and stagnant objects they were often taken to be. They were not like pebbles lying on the seashore, bits of matter that went nowhere unless pushed from behind by an outside force. To the contrary, there was some kind of inherent tendency among the world's fundamental particles to connect up in such a way as to bring forth first, simple orderly structures, then more complex ones, and then finally, the self-reproducing molecules of life itself.

BEGINNING WITH HIS stay at a cousin's house in Colorado, Dave Weininger entered his "season of exploration," a period of wandering around America and the world, searching for direction and purpose. And why not? This was, after all, the tail end of the sixties.

In 1970 he came back to New York State, and even though he hadn't graduated from high school and hadn't bothered with an equivalency degree or a diploma of any type, he managed to enroll at the Eastman School of Music, part of the University of Rochester. Music was one of the great loves of his life (the others being food, wine, and sex), and it was a "pure" activity, unsullied by political considerations such as the Vietnam War or anything else. It would suit him for the time being.

In 1972, however, he came to his senses and realized that chemistry, his first love, was a far better career choice than music. But because this was still the Vietnam War era and he was still very much anti–United States because of American participation in the war, he left the country for England and entered the College of Chemistry at the University of Bristol. While he admired the British tutorial system and ended up learning a lot of chemistry there, Weininger finally got tired of England and left that university too, without taking a degree. One high school and two universities, but still neither diploma nor degree to his name

So now, like the prodigal son, he came back to America and returned to Schenectady, where he took a job with General Electric and worked in their research and development lab as an assistant chemist, just as if he had never left.

That too got old very quickly. Dave departed once again, this time ending up at the University of Wisconsin. From there, in 1978, he got the first and only academic degree he would ever receive, a Ph.D. in civil and environmental engineering, with specialization in water chemistry. He now saw it as his mission in life to help rid the world's water systems of toxic wastes.

Wisconsin being covered with lakes (it is bordered by two of the Great Lakes, Lake Superior and Lake Michigan, and there are an additional fifteen thousand lakes inside the state, plus fifteen hundred streams), there was no better way to do field research on its vast and multiple bodies of water than by flying directly from one to another. Since the university owned a small fleet of Cessna light planes and had a flying program, Weininger naturally learned how to fly. Soon he was beholding the state's endless lakes, rivers, and streams from what he regarded as a godlike perspective.

The same year he got his degree, he took a job with the Environmental Protection Agency (EPA), working at the National Water Quality Laboratory in Duluth, Minnesota. A major part of his responsibility there was to catalogue all the toxic substances found in the various bodies of water in his jurisdiction. This was easier said than done, as the list of noxious chemicals soon ran into the hundreds, and many of them were unfamiliar compounds with complex names a mile long. The more normal pollutants were easy enough to record: for example, sulfur trioxide, nitric acid, and hydrogen peroxide, which were often found in smog. But the region's rivers and lakes harbored far more exotic stuff than that, as Weininger soon learned to his dismay, and there was simply no good way of keeping track of these more complex compounds by using traditional chemical nomenclatures, especially if you wanted to keep your records in a computerized database, which he did.

In late-twentieth-century chemistry, there were three main ways of representing chemical compounds, and all of them had substantial drawbacks from one point of view or another. To begin with, there was the systematic name of a compound, the way it would be designated in ordinary natural language: aspirin, acetone, glucose. Such nomenclature was good enough for simple compounds, but the names for even slightly more complex ones got out of hand in short order, as for example was the case with the names 3-(*para*-hydroxyphenyl)-2-butanone, or 2-methoxy-5-methylpyrazine, or thiopropionaldehyde-S-oxide. Complicated as they were, all those phrases were expressed in a form of "English," and they were not easily translatable into languages with different character sets, such as Russian, Greek, and Japanese. The systematic name of a chemical compound, then, was both excessively cumbrous and not truly generalizable to the world's other natural languages.

The second way of identifying a chemical compound was by means of its molecular formula, which itself could be either simple—for example, H_2O (water), $NaCl$ (salt), and H_2SO_4 (sulfuric acid)—or complex: for example, $O_2CC_6H_4CO_2C_2H_4$, which is called Dacron in the United States, Trevira in Germany, and both Terylene and Crimplene in the United Kingdom. While such designations were more universal than systematic names, chemical formulas nevertheless masked a hidden ambiguity, for although the formula listed the compound's chemical elements and their relative abundances within the molecule, it was silent as to the compound's molecular structure. In point of fact, one and the same formula might apply to two or more different substances depending on how their constituent atoms were physically arranged in the molecule: C_2H_6O, for example, was both ethanol (with the three elements bonded in one structural arrangement) and dimethyl ether (in which the same elements were arranged differently). Appearances to the contrary, a chemical formula did not always identify a unique molecule; in fact, one and the same formula often designated two or more compounds having substantially different structures and, therefore, different chemical properties. Put another way, completely different chemicals often shared one and the same chemical formula.

That problem was eliminated in the third conventional way of representing chemical compounds, by means of a diagram showing the compound's precise molecular configuration. Water, for example, was

Such a picture had the virtue of designating any given molecule uniquely: only one substance, water, looked exactly like that. And from the point of view of universality, the diagram had the dual advantages of being instantly intelligible to all the world's chemists and of looking exactly the same no matter what their native language happened to be. A picture was a picture.

The drawback to such pictures from Dave Weininger's point of view, however, was the impossibility of entering them into computers. There was no easy way to type such pictorial structures into any computer database that he was familiar with, especially when the molecules themselves, and the resulting diagrams, got complicated.

How, then, was he to do his job? How was he to create a master inventory of waterborne toxic substances if there was no efficient, universal, computer-compatible language in which to do so?

Above and beyond that narrow practical problem lay Dave's considerable agita, going back to childhood, over the lack of a truly universal, exact, language-independent, computer-friendly nomenclature in which the names and structures of all the world's many and diverse chemical compounds could be expressed. Chemistry was the most practical, fundamental, and all-encompassing science in the world—and in Dave's private view it was also the most beautiful—but there was as yet no ideal language for its substances.

He, therefore, would invent one.

4

SMILES

IN HIS ATTEMPT to create an ideal language for chemistry, Dave Weininger came at the end of a long line of like-minded hopefuls. Throughout the two-hundred-year history of the science, chemists had tried time and again to develop both a logical nomenclature and an all-encompassing yet easy-to-remember symbolism, and had achieved varying degrees of success. As a science, chemistry had arisen out of alchemy, and in consequence its early doctrines had incorporated a vast store of metaphysical baggage and mumbo jumbo, especially when it came to the matter of terminology. Prior to 1750, when chemistry began to emerge as a science, the alchemical and chemical texts that then existed were marked by the absence of any common chemical language or system of classifying or naming their objects, with the result that no chemist could be sure exactly what another one was talking about. Compounds were named in a variety of languages including Greek, Hebrew, Latin, and Arabic, along with a smattering of private and often even secret word coinages. Moreover, when alchemists bestowed names on things, they did so on no consistent principle but might name one substance for its color (*Spanish green* for copper acetate), another for where they imagined it came from (*Aquila coelestis* for ammonia), and still others for their smell, taste, texture, or other properties. In addition, the early chemists occasionally employed flowery metaphorical language to denote certain substances (*father* and *mother* for sulfur and mercury) as well as to describe the various types of chemical reactions (which they referred to as "gestations").

Taxonomically, in other words, the field was a mess.

Matters stood quite differently in botany and zoology, whose nomenclature had been brought to a state approaching perfection in the mid-1700s by the Swedish botanist Karl Linnaeus. Linnaeus, whose "passion for classification amounted almost to a disease," in the words of one writer, named the multifarious kinds of living organisms according to well-defined and rigorous principles. He drew up rules for fixing the limits of species and showed how each species differed from others. He developed a system of binomial nomenclature according to which each type of living thing was first given a generic designation placing it within the broader group to which it belonged, and then was differentiated from the other members within the group by means of a specific name that referred to, and only to, the plant or animal species in question. Genera in turn were grouped into classes, classes into orders, and so on, forming a system of successively nested divisions and subdivisions that was so elegant and well organized that it appeared to have been established by God.

The early chemists, by contrast, could only look on the Linnean system with envy and try to emulate it, which a few of them did. Between 1773 and 1775 Torbern Bergman, who had been a student of Linnaeus at the University of Uppsala, began applying the Linnean binomial principles to chemical compounds, proposing such names as *argentum nitratum* for silver nitrate ($AgNO_3$), *cuprum salitum* for copper chloride ($CuCl_2$), and *calx aerata* for calcium carbonate ($CaCO_3$), and the like. Bergman, however, was a mineralogist and limited his nomenclature chiefly to minerals.

In the 1780s a true Linnaeus of chemistry appeared on the scene in the person of Antoine Lavoisier. Lavoisier, from a well-to-do family and trained as a lawyer, had initially dabbled in astronomy, later took an interest in geology, and finally became an independent researcher in chemistry. He financed his investigations out of income generated by a half-million-franc investment in Ferme Générale, a private company hired by the French government to collect taxes. Ferme Générale was essentially a collection agency, but it trafficked in unpaid taxes rather than bad debts and as one would imagine, was not much beloved by the public. Lavoisier worked for the company as an administrator. In 1771, at the age of twenty-eight, he committed the additional sin of marrying the boss's daughter, a fabulously beautiful and highly intelligent girl who was exactly half his age. Marie-Anne Pierette Lavoisier had a talent for drawing and helped him with chemical illustrations. Earlier, at the age of twenty-three, Lavoisier had blackballed Jean-Paul Marat's application for membership in the French Academy of Sciences on the grounds that Marat was a fraud as a scientist. Marat, the French revolutionary, never forgot the affront.

For all his considerable image problems, Lavoisier nevertheless went on to become the father of modern chemistry and made a number of important discoveries. For example, he showed that air was made up of two main substances, oxygen and nitrogen. He also disproved the existence of "phlogiston," the substance that supposedly made combustion possible.

He was a zealot about the accuracy of chemical measurements, and he had a comparable monomania about the correctness of scientific terminology, which he wanted to be as precise as algebra. "We think only through the medium of words," he said. "Languages are true analytical methods. Algebra, which is adapted to its purpose in every species of expression, in the most simple, most exact, and best manner possible, is at the same time a language and an analytical method. The art of reasoning is nothing more than a language well arranged." He therefore decided to create a language for chemistry that would exhibit the desired rationality and precision.

Together with his colleagues Guyton de Morveau, Claude-Louis Berthollet, and Antoine-François de Fourcroy, Lavoisier set out a system of chemical nomenclature based on three basic principles: one, a given chemical substance ought to have a single fixed name that applied to it and it alone; two, the name ought to reflect the chemical composition of the substance, if it was known, but otherwise be noncommittal; three, the names should derive from Greek and Latin roots, while also being euphonious with the French language. In 1787, Lavoisier and the others published a three-hundred-page *Méthode de nomenclature chimique*, a system that was so clear and logical that it soon swept the field clear of other contenders and placed the science of chemistry on a new and rational foundation. In fact, with scant change, it is the systematic nomenclature in general use today.

Lavoisier's efforts were not appreciated by his countrymen, however. Apprehended in the wake of the French Revolution for his position in Ferme Générale, Lavoisier claimed that he was in fact a scientist, not a tax collector. Supposedly, the arresting officer replied, "The republic has no need of scientists," and on May 8, 1794, Lavoisier was guillotined and buried in an unmarked grave. "A moment was all that was necessary to strike off his head," Joseph-Louis Lagrange, the astronomer and mathematician, said at the time, "and probably a hundred years will not be sufficient to produce another like it."

In addition to providing an exact terminology, Lavoisier's *Méthode de nomenclature chimique* had included a proposal by Jean Henri Hassenfratz, the French inspector of mines, and Pierre Adet, of the Paris faculty of medicine, for representing chemical compounds by symbols instead of words. Their idea was that simple substances would be denoted by simple symbols while compound substances would be designated by combinations of them. Nothing could be more logical.

Unfortunately, the specific system they proposed was based not on the ordinary letters of the alphabet but rather on a system of newly invented geometrical patterns. The thirty-three chemical elements known at that time would be denoted by straight lines set at various angles. Metals would be circles, alkalis would be triangles, and so on throughout the list. Challenging enough to learn and remember ("more effort was required to remember their meanings than was saved by writing the symbols," said one historian of chemistry), the system posed an additional problem for typesetters and printers, who would need whole new cases of type in order to reproduce these cryptic signs and symbols on the page.

The Hassenfratz–Adet graphical scheme never caught on, but the notion of reducing the chemical compounds to a few well-chosen signs and symbols whose simple rearrangements could accommodate all the world's substances, known and unknown, was irresistible to later chemists. John Dalton, author of the modern atomic theory, advanced a pictorial system of his own in *A New System of Chemical Philosophy*, published in 1808. Each atom, he said, would be represented "by a small circle, with some distinctive mark [to identify the element]; and the combinations consist in the juxta-position of two or more of these." But Dalton's hieroglyphs, simple as they seemed at first, posed large obstacles to the human memory, not to mention to typesetters, and his system too died a swift death.

Ordinary language, and the typesetting boxes of printers and engravers, already had perfectly good symbols in the form of the English alphabet and Arabic numerals. Why not let them stand for the chemical elements and their relative abundances in a given molecule?

That was the proposal of Jöns Jakob Berzelius, the Swedish chemist who in 1813 claimed that "the chemical signs ought to be letters, for the greater facility of writing, and not to disfigure a printed book." Compounds, in his scheme, would be named for their constituent elements, using the initial letter (or, in cases of duplication, the first two letters) of the Latin name for each element, for example:

C = carbonicum
Co = cobaltum
O = oxygen
Os = osmium

The letters would be strung together as necessary to indicate the chemical makeup of a given compound, whereas numerals would be used as superscripts (later, subscripts) to show the number of atoms of each element in the molecule: S_2O_3 (hyposulfuric acid), for example, contained two atoms of sulfur and three of oxygen.

The system was a boon to printers and did not unduly tax the memory. Still, the notation had the signal disadvantage of failing to indicate the physical structure of the molecule in question, a weakness that was recognized at once by John Dalton, whose system of tiny pictures, for all its faults, nevertheless managed to show each atom's respective positions within the whole.

"Berzelius's symbols are horrifying," said Dalton. "A young student in chemistry might as soon learn Hebrew as make himself acquainted with them. They appear like a chaos of atoms. Why not put them together in some sort of order?"

That, indeed, was the question. What chemists wanted in their formulas was the maximum amount of structural and compositional information communicated in the smallest possible stretch of easily intelligible type. Molecules were complex three-dimensional objects, and the physical arrangements of their constituent atoms were all-important clues to how the compound behaved in chemical reactions. Knowing a molecule's constituents without knowing how they were ordered with respect to each other was like knowing the ingredients of a cake without knowing the proportions in which they were to be used in the recipe.

Fifty years after he elaborated it, the system of Berzelius still had not carried the day. In fact, so many competing systems had arisen in the interim that in 1866 there were at least twenty different formulas in the chemical literature for so simple a substance as acetic acid ($C_2H_4O_2$), the acidic constituent of vinegar.

The general problem, indeed, was rather daunting, for the challenge was to express, in a one-dimensional line of type, the complete atomic bonding patterns of a three-dimensional chemical molecule—the kind of information found (in two-dimensional form) in a typical structural diagram, as for example that of benzene:

Impossible as it might have seemed, a few chemists persisted in thinking that it ought to be possible to encode all of this structural, bonding, and other data in a single and solitary line of type. Many decades would pass, however, before the first plausible candidates appeared.

BY THE MID to late twentieth century, an additional constraint had been placed on any candidate system of chemical symbolism, for the omnipresence of the typewriter and the computer had made it necessary that a new notation make use of only those characters and symbols already available on a standard typewriter or computer keyboard. No system that made use of additional symbols would have the remotest chance of succeeding on a large scale.

Because the goal was to reduce the structure and atomic constituents of a molecule to a single line of type, the resulting system of nomenclature was referred to as a *line notation*. In theory, so efficient would be its encoding rules, and so compact the result of adhering to them, that a formula expressed in a successful chemical line notation would amount to no more and no less than a compressed diagram, a drawing expressed as a character string.

In the 1950s, one William J. Wiswesser, head of the chemical research department of the Ray-O-Vac Corporation, offered up a candidate system. It eventually became known as Wiswesser Line Notation, abbreviated WLN. In Wiswesser's line notation system, the allowable symbols were the letters A through Z, the numerals 0 through 9 (with the zero having an overprinted backslash to distinguish it from the letter O), a space as produced by the typewriter spacebar, and symbols such as the ampersand (&), hyphen (-), and the asterisk (*), all of which were found on the keys of any standard typewriter. Unlike Berzelius's system, however, the uppercase letters of WLN did not necessarily represent the elements, and the numbers did not represent the numbers of atoms within the molecule. Rather, Wiswesser's notation was *group-based*, so that letters, numbers, and combinations of them stood for various common molecular fragments or chemical groupings. For example, the number *1* represented the methyl group (CH_3-), the number *4* stood for the butyl group (C_4H_9-), and so on. Moreover, the various group symbols fitted together in such a way as to reproduce a whole chemical molecule, much as the pieces of a jigsaw puzzle fitted together to form the picture. For example, since *2* designated the ethyl group (C_2H_5-) while *Q* represented the hydroxyl group ($-OH$), the expression *2Q* stood for ethyl alcohol (C_2H_6O). Various other characters expressed structural information, such as whether the molecule was chained, branching, or formed a closed ring (as in the benzene molecule).

WLN provided chemists with unprecedented amounts of information about any given molecule, and by 1960 it had become the basis for the listing of chemical structures in *Chemical Abstracts*, a publication of the American Chemical Society. In addition, the notation had been adopted by a number of pharmaceutical and agricultural companies and by the U.S. Army. By 1985, some fifty different organizations held more than 3 million WLN records in computer storage.

Still, Wiswesser's notation had severe drawbacks that limited its ultimate usefulness to chemists. Because the system was based on coded groups, it was hard for a human user to read the formula and imagine the molecule; in other words, it was not user-friendly. More important, the task of learning the system was in itself extremely time-consuming inasmuch as the total list of its rules and regulations was laid out not in a technical paper, small pamphlet, or even a large one. In fact, the principles of the complete WLN system took an entire book to define, Wiswesser's *A Line-Formula Chemical Notation*, published by Thomas Crowell in 1954.

Perhaps most important, some of the information imparted by WLN was not contained *in the line* at all but appeared elsewhere on the page. Data on certain types of atomic bonds, for example, were to be found only in "connection tables," tabular listings that appeared separately from the line formula and often ran to a thousand or more characters for a single chemical compound. Wiswesser's system, in short, was not a line notation in the true sense of the word since some of its information burden had been off-loaded to places other than and apart from the line itself.

Single lines of type, apparently, could be packed with only so much chemical information and no more. So in 1981 when Dave Weininger decided to create a new, computer-based line notation essentially for his own private use, his prospects for success were fairly bleak.

Nonetheless, this was a man who had been thinking about chemical notation systems since early childhood, and several ideas were already rumbling around in his head. Essentially over the course of one long night in his apartment in Duluth, Weininger came up with a basic chemical line notation that consisted of just four short rules:

1. Atoms are represented by the conventional atomic symbols (i.e., those of Berzelius: C, N, O, Cl, Br, and so on).
2. Double bonds are shown by putting an equals sign (=) between the relevant symbols; triple bonds, by the pound symbol (#). For example, C=O represents the double bond between the carbon and oxygen atoms in the formaldehyde molecule:

3. Branching of a molecule is indicated by placing the branched atom or atomic grouping within parentheses. For example, C(=O) shows that an oxygen atom branches off from a carbon atom in the carboxyl group:

4. Ring closures are indicated by pairs of matching digits. For example, the two *1*'s in C1CCCCC1 (cyclohexane) means that the six carbon atoms are arranged in a closed ring:

Those four rules constituted the whole of Dave Weininger's initial line notation for chemistry. The application of the rules, however, was governed by few conventions. For one, hydrogen atoms did not have to be formally spelled out in the notation, but were understood to appear as they normally would in a given molecule according to the well-known laws of chemical valence. The symbol C, for example, stood not for a single carbon atom but rather for methane (CH_4), because a carbon atom, left to itself, automatically bonded with four hydrogen atoms. (To designate a carbon atom standing alone, the symbol C was to be placed within square brackets: [C].)

The system was almost too simple. Anyone literate in chemistry could memorize and learn to apply it within the space of an hour or so. And the more Weininger tested it, the more his system proved able to symbolize an extremely wide variety of chemical structures.

Still, a few of the more complex molecules defeated Weininger's four encoding rules, and for these he added a couple of extra principles as well as a few more governing conventions that pertained to special types of bonds, aromatic compounds, and other exotica. But that seemed to be about the end of it, and with this expanded system Weininger found that he was able to specify in a single line the chemical structure of practically any molecule with complete fidelity to the original.

Some of the resulting expressions, admittedly, proved to be lengthy and cryptic. Morphine, for example, which had a rather long-winded formula even in the Berzelius system $(C_{17}H_{19}O_3N)$, became, in Weininger's notation:

$$O1C2C(O)C=CC3C2(C4)c5c1c(O)ccc5CC3N(C)C4$$

Nevertheless, it worked!

Other chemists, however, were not convinced by Weininger's claims to have reduced the entire chemical universe to a simple line notation that could be fully stated in less than ten pages. Their usual response, whenever he told his friends and colleagues about the comprehensive but simple nomenclature that he'd developed for all of chemistry, was to smile benignly, as if he were crazy. So he called his system SMILES, which formally stood for *Simplified Molecular Input Line Entry System*, and called any given expression within it a SMILES.

Weininger now decided that with this he had invented a truly universal chemical notation, one with which the actual molecular structure of any arbitrary compound could be entered into a computer database by anyone who knew how to type.

AFTER LANGUISHING IN concept form on paper for three years while Murray Gell-Mann and George Cowan rounded up funding in the form of grants from the Carnegie Foundation, the MacArthur Foundation, and IBM (which also kicked in some free computers), the Santa Fe Institute formally opened for business in February 1987. For an organization designed to foster advanced, space-age work in chaos, nonlinear dynamics, and complexity theory (Gell-Mann's "plectics" terminology had been deep-sixed by everyone else), it was perhaps fitting that the building, at 1120 Canyon Road in the heart of the city's art district, exuded an air of permanence, stability, calm, and tradition.

Prior to its acquisition by the Santa Fe Institute, the old adobe structure had been a nunnery, the Cristo Rey Convent. It was a charming place, with ceilings held up by exposed vigas, rough-hewn fir beams that for all anyone knew had been carried down from the mountains on the backs of the ancient Anasazi. The bedrooms of the former nuns had been con-verted into offices, which tended to be on the smallish side, but some of the windows afforded views of the Sangre de Cristo Mountains just beyond. George Cowan, the institute's president, was installed in the office of the mother superior. The conference room, in the former chapel, was a private sanctum into which sunlight filtered through stained-glass windows, so that the enclosed space still felt halfway like a church. Cramped as it all was, and as otherworldly, the institute was nevertheless smack in the middle of historic, stylish, and expensive Santa Fe.

Santa Fe, called "the city different" by some residents, or at least by the city government and the Chamber of Commerce, had the aura of being sequestered from the outside world. It was extremely hard to get to, with only a small, out-of-the-way airport that had no regularly scheduled jet service; most visitors arrived by car. The city was an hour's drive north from Albuquerque (which was itself accessible by air only from Denver and a few other points), and the last stretch of it was up a grade known to the locals as *La Bajada* ("the lowering"), a steep ascent to seven thousand feet. There, at an elevation that made for an immediate shortness of breath, the traveler emerged onto a plateau that seemed to exist in its own special niche in space and time.

The sensation was explained by several factors, the first of which was its natural setting. The city lay on a mesa, a relatively flat stretch of land bor-dered on one side by the Sangre de Cristo Mountains, which rose to 12,600 feet, and on the other by the flat desert of the Rio Grande Valley. Across the valley to the west, some thirty miles away, was another range, the Jemez Mountains, which ran parallel to the Sangre de Cristos. Santa Fe, on the brow of a mesa halfway between flat desert and snow-capped mountain ranges, appeared to be suspended in space.

Then there was the color scheme, which was marked by three main hues, the primary one being pink. That was the color of the desert sands as well as the adobe buildings, which ran from light rose to rust brown. The darker hue was also the tone of the surrounding cliff faces, especially at sunrise or sunset, when all the city's colors tended to merge toward red.

There was the blue of the sky. Santa Fe weather was mostly sunny no matter what the season, the city was far from any source of industrial pollution, and the altitude made for exceptional visibility. Eighty miles was a relatively average straight-line seeing distance in the area, and the sky above was generally a deep, piercing blue.

The third color was green, which ranged from sage, the light grayish green of the sagebrush found all over the high desert, to the deeper green of the mountain pines, to the intermediate hue of turquoise. If there was any one item that characterized Santa Fe, it was turquoise, a semiprecious stone that could be found in every imaginable sterling-silver setting: on belt buckles, bracelets, necklaces, rings, watchbands, hairpins, hat pins, lapel pins, tiepins, cuff links, earrings, buttons, beads, money clips, key chains, faucet handles, doorknobs, dog collars, saddle, bridle, or stirrup ornaments, gun grips, holster snaps—this was the West, after all. In Santa Fe you could hardly look in any direction without coming face to face with one or more pieces of turquoise.

And then, finally, there was the culture, which gave the city the scent, flavor, and feel of a foreign land. The Spanish had founded Santa Fe ("Holy Faith") in 1607, thirteen years before the Pilgrims landed at Plymouth Rock, and there were Spanish names all over the place: town names like Cordova, Madrid, Pojoaque, and Española; street names like Cerrillos, Acequia Madre, Paseo de Peralta, and Camino Cruz Blanca. There were Spanish-named churches, schools, hotels, markets, restaurants, and fiestas. Newcomers to the area soon adjusted to the language and even came up with some reasonably Spanish-sounding names of their own. Stan Ulam and John von Neumann, the Manhattan Project scientists, dubbed the local police station "El Palacio de Securita" and renamed a Los Alamos church "San Giovanni delle Bombe."

But in fact the Spanish were not the primary linguistic, visual, or cultural force of Santa Fe. That distinction was held by the American Indian, who had lived in the area more than a thousand years before the Spanish arrived and whose images, artifacts, costumes, designs, folklore, and in some cases, even the food—blue corn, posole (a hominy and bean stew), and fry bread—dominated the scene. The city was surrounded by Indian pueblos: Pecos Pueblo, San Juan Pueblo, San Ildefonso Pueblo, Santo Domingo Pueblo, San Felipe Pueblo, the pueblos at Nambe, Jemez, Zia, and others. Native American artifacts—pottery, baskets, jewelry, rugs, blankets—were for sale everywhere in the city.

The town itself was a rich mix of hotels, shops, coffeehouses, restaurants, art museums, and art galleries. Santa Fe residents, some of them, took an inordinate pride in the fact that the city's art galleries, of which there were 150, and Indian shops, another fifty, vastly outnumbered such more mundane and practical businesses as supermarkets, convenience stores, and gas stations. This was a town where people placed great reliance on pyramids, crystals, flying-saucer music, massages, and mystic cults that had originated somewhere in the Gobi Desert. It was a city, residents said, of sixty thousand people and seventy thousand religions.

The center of it all was the downtown Plaza, a small square that itself seemed set off from the town that surrounded it. Long ago, the square had marked the western end of the Santa Fe Trail, a westward migration route that began at Independence, Missouri, and continued on for eight hundred miles. Today's Plaza consists of a small grassy park bordered on three sides by two-story buildings and on the fourth by the Palace of the Governors, a long, low, single-story building that looked like a stage set for a cowboy movie but was in fact wholly original. Built by the Spanish in 1610, the Palace of the Governors was America's oldest public building in continuous use. Along the facade was a block-long, porchlike wooden colonnade, a covered walkway where on any given day in summer or winter, Indians in native dress would spread out their wares: yet more pottery, silver, and turquoise, together with dolls, blankets, clay pipes, feathered headdresses, and homemade Indian moccasins and other apparel. With mesquite smoke rising into the cool, dry air from the sidewalk barbecues, with the bells of St. Francis Cathedral tolling the hours, and with the pine-scented mountains rising into the blue, the Plaza somehow seemed entirely disconnected from the rest of the United States.

A really snappy dresser in Santa Fe would look like a Martian on the streets of New York, Washington, or even San Francisco. This person would be in a gray felt ten-gallon hat ringed by a black leather stampede strap; a white pleated shirt and bolo tie with a turquoise sunburst tie-slide from Shalako Traders on the Plaza; a tan buckskin jacket with matching leather fringes across the chest, back, and along the sleeves; blue jeans;

and—of course—a pair of powder-blue brushed-elk cowboy boots from Back at the Ranch, at the corner of Marcy and Otero.

For divertissement, the genuine Santa Fean did not go to a basketball game, football game, or God forbid, baseball game, but rather went to the rodeo, hiked to the peak of Santa Fe Baldy, or went horseback riding, and then on the way back stopped at Pagosa Springs for a hot springs soak or at 10,000 Waves for a Japanese hot stone massage.

Yes, this was a different world, all right.

That a new high-tech scientific research center, the Santa Fe Institute, should put down its roots here was, nevertheless, entirely appropriate. The institute's researchers, after all, would be charged with the task of creating wholly new sciences, novel ways of perceiving nature, offbeat theories, new paradigms, disciplines that never before existed anywhere in the world. There was a feeling of vastness in Santa Fe, an openness: there was more air, more sky, more sun, more space, more time—all of which encouraged intellectual boldness and radical thoughts.

Soon the institute would be brimming over with these newly minted theoretical artifacts, thought structures previously unknown to humankind: genetic algorithms, artificial life-forms, nonequilibrium systems, fitness landscapes. Whatever they were, and whatever they meant, the blue New Mexico sky equally and impartially embraced them all.

IN NOVEMBER 1985, prior to its formal opening in the Canyon Road convent, and when it was still just an organization that existed only on paper, the Santa Fe Institute sponsored its first scientific conference, a workshop on superstring theory. Superstrings were the new-wave theoretical primitives of particle physics—invisible wisps of energy throbbing and pulsating at the innermost core of matter. But for all their exoticism they were still just the standard fare of conventional reductionistic physics, and there was nothing really special about them insofar as the institute's overall mission was concerned. Elementary particle theory was not exactly what the Santa Fe Institute was all about, but Murray Gell-Mann, chairman of the board, had nevertheless wanted a superstring conference, and he got it.

"The president of the institute, George Cowan, hated these things bitterly," Gell-Mann said of superstrings. "I don't know why."

Later, in August 1986, the institute sponsored another meeting in Santa
Fe, this one on the subject of the global economy. Economics not being a
science, it too was not primarily what the institute had been created to
explore. On the other hand, the world economy was arguably a dynamical
system, perhaps even a chaotic system, its workings often subject to wild
perturbations provoked by essentially trivial (if even perceptible) stimuli.
If there was a statutory-defining characteristic of a chaotic system, it was
captured by the phrase *sensitive dependence on initial conditions*—the
famed "butterfly effect," whereby an insect flapping its wings in Brazil
later caused typhoons in China. Decidedly, economic phenomena—stock
market trends, oil prices, interest rates, banking crises—seemed to behave
that way all too often, open to violent mood swings, many of them precipitated by next to nothing.

So the institute held an economics workshop at Rancho Encantado, a
dude ranch north of the city, bringing together Cowan, some of the institute's staff members, and John Reed, the incoming chief executive officer
of Citicorp, the New York bank that had recently lost $1 billion in bad
loans to Third World countries, with the prospect of $13 billion more also
evaporating into thin air. If chaos theory could help to avert such debacles
in the future, well then John Reed was all for it.

Whether the workshop itself had more than a butterfly wing's worth of
impact on the subsequent course of the global financial system was debat-
able, but the event was nevertheless a milestone in the institute's history
and set the stage for an ambitious economics program later to be developed at the institute.

As it was, it was not until July 1987, after the Santa Fe Institute had
been at the Canyon Road location for six months, that it hosted the type
of event that fully answered the purpose for which it had been founded.
The conference title alone hinted that a radically new theoretical construct was about to be born, for the event in question was called the
"Matrix of Biological Knowledge Workshop."

As a science, biology had long been characterized by an extreme frag-
mentation and a corresponding lack of general unifying principles such as
were found, for example, in physics. "Physics has a few basic principles, a
few elementary particles," said Harold Morowitz, a professor of biology at
Yale and a principal organizer of the workshop. (Back in their early days
as Yale undergraduates, Harold Morowitz and Murray Gell-Mann had

been lab-bench partners in sophomore physics.) "In biology you have three million species, you have hundreds of thousands of macromolecules, you have all the metabolic systems and so forth—you have total information overload."

Biology was spread across a variety of different levels of structure and organization, levels that ranged from molecular biology—beginning with protein structures and rising to the complexity of DNA—to the organelles inside cells, to individual cells themselves, to developing embryos, to the organs and tissues of higher organisms, to whole, fully formed organisms, to families of them, all the way up to complex ecological systems. No other science was characterized by a comparable degree of hierarchical stratification.

That multiplicity of levels, in turn, posed barriers to understanding, especially if you wanted to cross from one domain to another. For example, if you wanted to know how a mutation of a given molecule affected the behavior of an organism, you were asking a question at the molecular level that pertained to results at the behavioral science level. The problem was that there was no common language, theory, or conceptual structure that encompassed both domains: molecules spoke the language of enzymes, Brownian motion, and chemical bonding rules, whereas organisms existed in a wholly different universe of discourse. Moreover, there wasn't any database that allowed you to pose the molecular question and get the whole-organism behavioral answer, and such answers could be crucial in certain contexts. Surgeons who did liver transplants, for example, wouldn't necessarily be experts in molecular structure, but it would turn out in fact that tissue rejection occurred essentially at the molecular level.

The goal of the Matrix conference was to create a framework that would bridge the gaps and unite the extremes. The general idea was that biological knowledge constituted a vast matrix: a gigantic interconnected web of basic fact, experimental data, clinical results, and theory spread out across all the various levels of biology, from biochemistry to ecology. The final goal, which was to serve only as a regulative ideal rather than as a concrete objective to be achieved during the conference, was to pack all of that fact and theory into a single, unified database and organize it in such a way that any competent biological researcher could search through it at will. Further, and more portentously, a distant goal was to construct a database so rich and supple that it would yield new generalizations and biological laws—perhaps even a biological Theory of Everything.

All this, of course, required computers, for the Matrix conference would be an attempt to put all the world's biological knowledge into computers and make it easily available to the user. As Harold Morowitz would later claim, "Computer science is to biology what calculus is to physics. It's the natural mathematical technique that best maps the character of the subject."

Biological systems were complex systems in the highest degree, which meant that they were tailor-made for computer modeling. With sufficiently powerful computers, a deep enough database, and the right software, one ought to be able to construct simulations of all kinds of biological processes: metabolic events, the inner operations of bacteria, maybe even simulations of whole organisms.

"In principle," Morowitz said, "we'll be able to identify every enzymatic step that a given cell can do. We will then be able to do a complete computer model of the entire operation of the cell. Then we will be able to do experiments where we go in the lab and we give the cell a slug of sugar, and we go to the computer program and give it a slug of sugar. And if we've truly understood it, and gotten all the steps right, that would give us a kind of profound understanding, and it would have come from applying computers to biology."

That, in sum, was the ultimate hope, the final payoff of the Matrix of Biological Knowledge Workshop: reducing organisms to information, and then using that information to understand, predict the behavior of, and perhaps even repair everything from single cells to whole organisms and entire ecological systems. An ambitious goal by any standard, but there was nothing inherently impossible about it.

5

GRASP

THE MATRIX OF Biological Knowledge Workshop was held over five weeks in July and August 1987. Since the Santa Fe Institute's Canyon Road offices were barely large enough to contain its own staff, the workshop's faculty members, participants, and day-to-day activities were housed on the campus of St. John's College. St. John's, a private four-year liberal arts college devoted to the "great books" program ("where Plato is your professor," according to an account in *Smithsonian* magazine), was located on the side of a mountain above Santa Fe and afforded a fabulous view of the city and its surroundings.

Harold Morowitz and his colleague Temple Smith, of Harvard Medical School, had recruited a core faculty of sixteen professors from various disciplines to serve as lecturers and group leaders. Among them, there was Patrick Winston, from MIT's Artificial Intelligence Lab; Hans Bode, a cell biologist at the University of California; Robert Goldstein, a specialist in database management at the University of British Columbia; and David Garfinkel of the University of Pennsylvania, who was working on a computer simulation of cellular metabolism.

To get participants (not "students"—most of them already had doctorates), the two organizers placed ads in *Science* and *Nature*; sent announcements to major departments of biology, biochemistry, and computer science; and also spread the word by e-mail. The workshop was billed as a foundational, breakthrough attempt to apply computer technology to biology, thereby fundamentally advancing the progress of biological science.

There would be lectures in the morning and group sessions in the afternoon, and in the evenings a bunch of IBM PCs would be available for work and play, the whole enterprise to take place in the lofty atmosphere of the Sangre de Cristos high above Santa Fe. Furthermore, all expenses would be paid out of grants from the National Science Foundation, the National Institutes of Health, the U.S. Department of Energy, and the Alfred P. Sloan Foundation, who were supporting the effort under the auspices of the Santa Fe Institute.

Who could resist?

Certainly not Anthony Nicholls. At the time he first heard of the workshop, Nicholls was a doctoral candidate at the Institute of Molecular Biophysics at Florida State University (FSU) in Tallahassee. Later, in 1997, he would settle in Santa Fe permanently and start his own company, Open-Eye Scientific Software.

Nicholls had been born in Plymouth, England, in 1961 and originally wanted to be a physicist. He later developed an interest in biology, and so while he read physics at Oxford, he looked around for graduate programs in biophysics. He applied to a bunch of them in the United States and wound up at Florida State.

He started off in the lab doing immunology, which he thought looked like a promising research direction. He discovered soon enough, however, that he was not cut out for lab work, that he had no special aptitude for manipulating pipettes, filling hundreds of tiny wells in plastic trays with minute droplets of fluid, running electrophoretic gels, or any of the rest of it. What he was actually interested in was theory, and in particular the theory of how proteins worked.

Proteins were not exactly new to science. Emil Fischer, Dave Weininger's boyhood hero, had worked out the primary structure (the amino acid sequence) of a simple synthetic protein molecule by 1907. By 1957, Max Perutz and John Kendrew had worked out the complete structure of a far more complex natural protein, myoglobin. Still, proteins were the workhorses of the body, and almost every bodily component was made up of, affected by, or had some effect on various sets of proteins. Further, many protein molecules were hugely long and complex chemical entities. Why were these molecules so versatile? How did they possess the magical ability to do all and be all, all the time, everywhere?

The answer was that what made proteins so omnipresent, adaptable, and useful was the phenomenon of "information transduction," the ability of a protein molecule to convert information from one type to another. For example, when a fragment of ATP (adenosine triphosphate, an energy-bearing molecule) bumped into a protein, it attached itself to the protein, whereupon the latter molecule suddenly changed its shape in response, as if it were reforming itself to accommodate the successful attachment of the ATP. The question was, what were the mechanisms that enabled the protein molecule to do that?

While Anthony Nicholls was still a grad student, one theory had it that impulses traveled around proteins in the form of *solitons*, isolated waves that propagated through matter. The energy of the ATP molecule was thought to be about twice that of the vibration of a carbon-oxygen bond, of which there were many in protein molecules, so the idea was that the energy from the ATP would pass through the protein by means of successive vibrations of the molecule's carbon-oxygen bonds.

As part of his doctoral work, Nicholls wrote a computer program to simulate the workings of solitons for the purpose of finding out whether they could in fact explain information transduction in proteins. He wrote the program in FORTRAN, an early scientific programming language, which he ran on an early supercomputer at FSU, a now extinct dinosaur of a machine called a Cyber 205, manufactured by Control Data Corporation. The Cyber 205 was enormous, so big that it almost filled a room.

Nicholls spent many hours writing his program, which in the end was so streamlined, succinct, and optimized that it ran in about a minute, an extremely fast-running time for the era. What his simulation showed was that there was a mismatch between how soliton waves were thought to propagate and how proteins changed their shapes when affected by other molecules. Solitons, therefore, were not the mechanism that explained information transduction in proteins. The structure of a protein molecule was the product of a delicate balance of forces among the molecule's constituent parts, and any interaction with another molecule upset that balance and precipitated a change in the global shape of the protein. Neither solitons nor any other single phenomenon explained the resultant energy transfers.

Nicholls had discovered in writing the program that his true calling was not experimental biology but rather theory, simulations, and the crafting of computer software that reproduced the processes of life—the very enterprise that he was now about to embark on in a professional research context at Harold Morowitz's Matrix of Biological Knowledge Workshop.

Nicholls arrived in Santa Fe in the second week of July 1987, together with Teresa Strzelecka, who was also a grad student at the Institute of Molecular Biophysics at Florida State. She and Nicholls would later get married.

New Mexico was a revelation to Anthony Nicholls.

"You have to understand that as an Englishman you grow up in a very tiny, claustrophobic kind of country, although you don't realize it at the time," he said later. "And you come out here and you're able to see, you know, 200 miles. In Britain, that is like the country—the entire country!

"It doesn't affect everybody that way," he added, "but if it gets into your blood, it never leaves."

George Cowan gave a brief welcoming talk at the start of the Matrix workshop, telling the assembled researchers that the institute's founders had originally thought of concentrating their efforts on the biological sciences and had even modeled the institute's constitution and bylaws on those of the Salk Institute and Rockefeller University, two of the nation's foremost centers of biological research. They had strayed from biology because John Reed, of Citicorp, had offered to support a major program in economics. The merging of biology and computers, Cowan said, nevertheless remained very much on his mind. His introduction was followed by a separate welcoming talk by Michael Riccards, the president of St. John's College, after which the workshop began in earnest.

Nicholls and Strzelecka wound up working on the metabolic map project, an attempt to construct a simulation of the major metabolic processes. Metabolism was a tremendously complex subject in biology, and the traditional metabolic charts hanging on the walls of biology labs looked like diagrams of all the world's subway systems combined into one. A simulation of metabolic reactions would be a boon to researchers, imparting a measure of needed intelligibility to a subject that was otherwise a hodgepodge.

The Santa Fe Institute had placed six of its IBM PCs in the college's chemistry lab, and over the five weeks of the workshop, the members of the metabolic map group packed as much information as they could, about metabolic compounds, reactions, metabolic pathways and suchlike, into a computer database. In the end their database included information on a total of 546 distinct metabolites that participated in 308 different reactions, and traced metabolic pathways for glycolysis, most amino acids, purines, and some lipids and steroids.

The group did not have enough time to go beyond that compilation and construct an actual working simulation of metabolic processes. Still, it was a start.

Other groups made similar progress toward constructing databases for macromolecules (such as proteins), growth factors, and peptides. In the end, everyone had produced a database, but nobody had constructed a working simulation.

Over the weeks a number of VIPs showed up, including Garrey Carruthers, the governor of New Mexico, and U.S. Senator Pete Domenici. One day, the entire group was loaded on a bus and driven to Los Alamos, where they got a tour of the lab and the Bradbury Science Museum. This was fitting, for Los Alamos had pioneered the use of computers to simulate physical processes. The fact that the lab's major product was bombs, however, clearly bothered Anthony Nicholls. Off in a corner, in its own display case, was a mock-up of a neutron bomb, presented as if it were a triumph of modern technology, a viewpoint that Nicholls did not share.

The Matrix of Biological Knowledge Workshop, little known beyond the small band of researchers who attended it, was a watershed event in the emerging marriage of computers and biology, and a formative influence in the birth of the Info Mesa.

AFTER THE MATRIX workshop, Anthony Nicholls and Teresa Strzelecka returned to Florida. They received their doctorates, got married, and then moved to New York, where they had secured postdoctoral research positions at Columbia University.

At Columbia, Nicholls found a position with Barry Honig, a biochemist who worked in the field of molecular electrostatics, the science of understanding, explaining, and calculating the electronic force fields around molecules. Knowing the shapes, sizes, and strengths of such force fields was useful in predicting a molecule's behavior in solvents as well as its ability to interact with other particles. It was of particular interest to understand the electrostatics of larger structures (called *macromolecules*) such as proteins and DNA, because much of a macromolecule's biochemical behavior was dependent on the contours of the electronic charges surrounding it.

In an attempt to calculate the electric potentials around macromolecules, Honig applied the Poisson-Boltzmann equation at the molecular level. The equation was a standard item in classical physics, but it was typically used for calculating the forces pertaining to larger phenomena such as parallel-plate capacitors and the like. Honig thought he might be able to utilize the same formula at the molecular level by developing an appropriate computer implementation, and in the end he succeeded in writing a program that solved the Poisson-Boltzmann equation for a range of macromolecular objects, including proteins.

He called the program DelPhi. Each calculation using the DelPhi program took about thirty minutes, and two separate calculations were required to get the full shape of the electric potential that characterized the molecule in question. The program ran on a Convex C2 computer, another relic, made by the Convex Computer Corporation of Plano, Texas, a firm that later became part of Hewlett-Packard.

It was at about this point that Anthony Nicholls arrived in Honig's lab. Having just recently finished his soliton simulation work, Nicholls took great interest in Honig's DelPhi application. But he thought it was unreasonably slow, and so with Honig's permission, Nicholls set out to rewrite the DelPhi code with a view to speeding it up somewhat.

In 1989 Nicholls rewrote the code from top to bottom and made it sixty times faster than the original. In fact, he got it to where the software calculated the answer in just a couple of minutes, providing the user with complete information about the electrostatics of any protein for which the user supplied the correct input data.

He called the improved program DelPhi II, and it quickly became a mainstay of biophysics. It was soon commercialized by BioSym, a software firm, under the trade name "DelPhi II, A Macromolecular Electrostatics Modelling Package." Nicholls started collecting royalties on it, receiving a check in the mail every February for a few thousand dollars. He deposited the amounts into a savings account and forgot about them. While he liked getting royalties, the downside of the arrangement was that nobody associated the DelPhi software with Anthony Nicholls, since all he'd done was to optimize the program—he hadn't actually created it.

Still, Nicholls saw an opportunity here. Useful as it was to protein chemists, drug developers, and the like, the DelPhi II program had the substantial drawback that the results it provided were strictly numerical, a long list of numbers printed on a page. It would be far more useful, Nicholls thought, if the output were visual. Proteins were famous for interacting with other molecules in lock-and-key, tongue-in-groove fashion, so if you were interested in understanding how a given protein interacted with its environment and with other bits of matter, there was no substitute for having in front of you a crisp visual picture of what the protein itself actually looked like. Theoretical understanding was all well and good, data was essential, numerical results could not be done without—all that was true. But a picture of the molecule you were interested in was worth a thousand numbers. There were already some programs that converted molecular numerical information into pictures, but the problem with them was that they took forever to run. Nicholls wanted to adapt DelPhi II so that the output would not be numbers but rather an image, and an image that would appear fast, within a minute or so of your inputting the data and perhaps even within a matter of seconds.

At about this time Honig's lab took delivery of an early Silicon Graphics workstation. It was a beautiful, compact machine: no more colossal room-filling mainframes—this was like having a supercomputer on your own desktop. The computer was fast, the screen resolution was superb, and more important, the whole thing seemed to be geared toward pictures and graphics.

Nicholls got hold of the user manual and turned to the graphics section to get an idea of how much work would be involved in turning numerical data into visual depictions of protein molecules. Everything he saw led him to be believe that he could do what he wanted to do without heroic amounts of labor.

But in addition to showing any given protein as a physical structure, he wanted his program to depict the interface between the protein molecule itself and the solvent that surrounded it, such as water. That interface was an electrostatic surface; it could be understood as an envelope covering the molecule like a cellophane shrink-wrap. The surface was important to protein chemists because it was what actually mediated the relationship between the molecule itself and its surroundings.

In short order Nicholls wrote a program that took a batch of numbers that characterized the component parts of a protein, and with it, generated a picture of the molecular structure to atom-by-atom resolution. The same program also created a second on-screen image that showed the electrostatic surface surrounding the molecule. Whereas the other available molecular modeling programs took literally hours to create an image from numerical data, the code that Nicholls had written took just a few seconds to produce an image of the molecule and no more than a fraction of a second to generate an image of the electrostatic shrink-wrap that surrounded the structure. The two images appeared side by side on the screen, so that the user could easily compare the two.

He named his program Grasp, for Graphical Representation and Analysis of Structural Properties. To run it, the user had to input the data that specified the locations of the atoms that made up the protein. That information was stored in the Protein Data Bank, maintained by the Brookhaven National Laboratory in Long Island. A protein was a complex object consisting of amino acids that were themselves composed of several atoms, and the numerical data that described the positions of those atoms often ran to hundreds of lines of text. The Protein Data Bank supplied the Cartesian coordinates specifying the X, Y, and Z spatial locations of each atom, plus certain other data points, for any given amino acid. For example, the atoms that made up the first two amino acids in a sample protein were listed as:

ATOM	1	N	ASP	L	1	4.060	7.307	5.186	1.00	51.58	1FDL	93
ATOM	2	CA	ASP	L	1	4.042	7.776	6.553	1.00	48.05	1FDL	94
ATOM	3	C	ASP	L	1	2.668	8.426	6.644	1.00	49.84	1FDL	95
ATOM	4	O	ASP	L	1	1.987	8.438	5.606	1.00	50.83	1FDL	96
ATOM	5	CB	ASP	L	1	5.090	8.827	6.797	1.00	50.57	1FDL	97
ATOM	6	CG	ASP	L	1	6.338	8.761	5.929	1.00	54.09	1FDL	98
ATOM	7	OD1	ASP	L	1	6.576	9.758	5.241	1.00	56.90	1FDL	99
ATOM	8	OD2	ASP	L	1	7.065	7.759	5.948	1.00	51.06	1FDL	100
ATOM	9	N	ILE	L	2	2.249	8.961	7.803	1.00	45.48	1FDL	101
ATOM	10	CA	ILE	L	2	0.920	9.547	7.949	1.00	38.04	1FDL	102
ATOM	11	C	ILE	L	2	0.950	11.039	7.634	1.00	39.85	1FDL	103
ATOM	12	O	ILE	L	2	1.800	11.770	8.153	1.00	39.76	1FDL	104
ATOM	13	CB	ILE	L	2	0.438	9.271	9.402	1.00	37.16	1FDL	105
ATOM	14	CG1	ILE	L	2	0.290	7.766	9.577	1.00	31.41	1FDL	106
ATOM	15	CG2	ILE	L	2	-0.884	9.974	9.690	1.00	34.89	1FDL	107
ATOM	16	CD1	ILE	L	2	0.141	7.273	11.009	1.00	32.83	1FDL	108

After inputting the data, the user would specify a few parameters, soon after which two vivid structures would appear on the display screen:

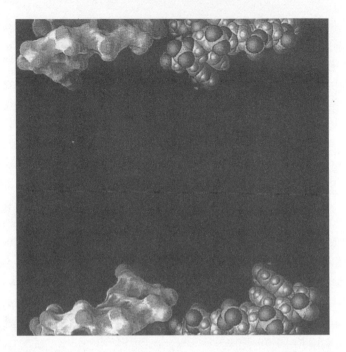

The different atoms were color-coded so that the viewer could tell which atom was which. Also, the program's images were depth-shaded to black so that individual atoms appeared darker the farther they were from the viewer, giving an illusion of depth.

Nicholls's software also included a variety of user-tunable functions, allowing the representation of atoms and bonds in "ball-and-stick" format instead of as spheres, and in other commonly used scientific formats. In addition to picturing the protein molecule, Grasp allowed the user to create an image of a drug molecule so long as the Cartesian coordinates of its atoms were known. Because it pictured each separate atom of the molecule, the program was particularly good at showing the various cavities, dents, or physical depressions where specific candidate drug molecules would be likely to attach themselves.

For a protein scientist or a drug developer, the ability to generate such images at will, in a matter of seconds, was nirvana. In fact, Grasp was so popular with scientists that it quickly became the default system for depicting the exterior structure of any new protein. Within a short time after he released the software, any protein molecule that was pictured in *Nature*, *Science*, or other scientific journals had almost invariably been produced using Grasp.

Unfortunately, Nicholls would derive no royalties from his invention. He had developed Grasp in Barry Honig's lab at Columbia University, and so it was Columbia, not Nicholls, who owned the program and reaped the benefits. The university licensed Grasp to both academic and industrial users for a flat fee of $500, payable directly to the university. None of it went to Nicholls himself.

STU KAUFFMAN WENDED an exceptionally circuitous route, intellectually, professionally, and personally, before ending up at his final destination in New Mexico. It was a course that was to take him first to Pennsylvania, then to the Los Alamos National Laboratory, and only afterward to Santa Fe, where he'd become a visiting professor at the Santa Fe Institute and the temporary nemesis of Murray Gell-Mann. Along the way, he would nail down his theories as to life's origins, the birth of order, and the emergence of complex systems in nature, about which he would publish two books, one technical and one popular. And even while at the height of all this abstract thought and theorizing, he had time for more practical pursuits such as inventing and patenting a new method for creating drug molecules, as well as starting a biotechnology company, Darwin Molecular, to capitalize on his invention. In 1987, finally, as a capstone to all this, he was awarded a MacArthur "genius" grant. A substantial track record for any scientist.

In time, this set of diverse interests and accomplishments would operate to place Stu Kauffman firmly on the Info Mesa, where he'd meet up with and become the colleague of Dave Weininger, Anthony Rippo, and Anthony Nicholls, all of whom would be running their own scientific software businesses within a few miles of Kauffman's near the downtown Plaza.

In 1975, however, all of this was still to come. That was the year in which Kauffman accepted an associate professorship in the department of biochemistry and biophysics at the University of Pennsylvania medical school in Philadelphia. Inner-city Chicago, he and his wife Elizabeth had decided, was not the ideal place to raise a family. So when the University of Pennsylvania offered Kauffman not only immediate tenure but also the run of the lab, he made the change. Stu was in fact not particularly good in the laboratory; he was not a born experimentalist by any means but wanted to learn the ways of biomolecules, cells, and larger organisms firsthand, and a biology lab was the place to do it. His later critics to the contrary, Kauffman was never one to stray very far from experimental fact.

Stu would spend the next twenty years at Penn, gradually realizing that the two major theoretical insights he'd had during the course of his professional life were now finally merging into one big theory. His first passion, genetic regulatory systems, was a mania that had gotten him into light-bulb networks and large-scale computer simulations. An elementary sort of order, he'd discovered through the simulations, could emerge quite naturally from simple and disorganized starting points. It was while working on the origin of life, his second big interest, that he'd come upon the notion of autocatalytic sets, networks of self-organizing and self-sustaining chemical chain reactions. Such networks, he had concluded, were mechanisms that could account for the development of spontaneous order in the real world, the very same sort of order that he'd first discovered back in his light-bulb days.

What Stu Kauffman was interested in was explaining how the world's order had arisen, and with his theory of self-catalyzing reactions, he seemed to have found an answer. He was exploring something very broad and deep, the almost metaphysical question of how the dead matter that made up the universe had managed to organize itself into crystals, cells, trees, animals, and entities such as himself.

By the early 1980s, Kauffman realized that he was not the only one in the world thinking about the subjects of order, networks, complexity, and emergent behavior. Soon in fact, there would be a "science of complexity" and an entire raft of "complexity theorists" bent on elaborating their own private versions of how emergent behavior and complex systems came to be, where they were going, and what it all meant. Kauffman therefore started writing his first book, *The Origins of Order*, setting forth his own view of how self-organizing systems arose in nature and how they were amplified, diversified, and elaborated across time by the standard processes of evolution by natural selection.

In 1982, shortly after he'd launched into the writing, Kauffman attended a scientific conference in Bavaria where he met one Doyne Farmer, a tall, thin, pony-tailed physicist who was then working at Los

Alamos. It was one of those chance encounters that seemed fated in retrospect—as if ordained by a cosmic alignment of the sun, planets, and moons of the solar system.

Doyne Farmer was a man with a past. Christened James Doyne Farmer Jr. by his parents, he went by his second name, which he pronounced "Doan," as in *Zen koan*. He'd grown up in Silver City, New Mexico, a fabled mining town in the foothills of the Gila National Forest, where he spent his high-school years overhauling motorcycle engines, fiddling with electronics parts, and building amateur rockets. From those promising antecedents, Farmer went to Stanford, where he majored in physics, after which he went to the University of California, Santa Cruz, for graduate work.

That is where he met up once again with his lifelong sidekick, Norman Packard. Packard, a somewhat ascetic-looking specimen with a lock of brown hair forever falling down over his eyes, was the cousin of David Packard, cofounder, with Bill Hewlett, of Hewlett-Packard, the computer company. Norman Packard, it turned out, although he had been born in Montana, had also grown up in Silver City, New Mexico. In fact, he and Doyne Farmer had been buddies at the time, inseparable, like twins. Doyne had even lived for a few months with the Packards, who owned the biggest house in town, an enormous place with twenty-three rooms. The house was so big that as a kid Norman ran two businesses from it, a tropical fish store on the ground floor ("The Silver Aquarium") and an electronics shop upstairs, where he did contract work for a local television repairman.

When Norman went away to college, he too majored in physics, earning his undergraduate degree at Reed College in Portland, Oregon, before ending up, like Doyne, at the University of California, Santa Cruz. There, the two of them joined forces at a higher level, tackling problems that were much in vogue then at the Santa Cruz physics department: long-range weather prediction, turbulence in the interstellar medium ("cosmic arrhythmias"), the movement of air across a wing, the passage of oil through a pipeline, the dripping patterns exhibited by a leaky water faucet. Most of these phenomena had two things in common: they were instances of fluid flow in one form or another, and they were examples of unpredictable, nondeterministic systems. In other words, they were classic nonlinear or "chaotic" phenomena.

Farmer, Packard, and two other Santa Cruz grad students, Rob Shaw and Jim Crutchfield, banded together informally to analyze such systems and called themselves the Dynamical Systems Collective, more informally known as the Chaos Cabal. They decided they were discovering the fundamental laws of chaotic phenomena, whether natural or artificial.

The group went on to do much of the foundational work in what later came to be known as *chaos theory*. Hallmark of their methodology was an extremely heavy reliance on computers. The phenomena under investigation were so complex, the underlying behavior patterns so subtle, fleeting, and hard to detect, that there was no other way of making progress than by simulating the systems with appropriate software. There was simply too much information to deal with, too great a volume of sheer data to be reduced to manageable order, for scientists to make any headway otherwise.

Doyne and Norman did not confine their computer simulations to such relatively dim and dull stuff as global weather patterns and dripping water faucets, however. The two of them had a knack for applying the farthest-out, most abstract scientific theories to more offbeat real-world problems, and at some point it occurred to them that gambling was as susceptible to chaos-theoretical treatment as any other supposedly "unpredictable" phenomenon. Outside of quantum physics, no phenomenon was truly unpredictable, not if you approached it properly. Even the gyrations of a roulette wheel, they decided, were grist for the mill of chaos theory. A roulette wheel, after all, was a self-contained chaotic system—it was like a miniplanetary system subject to the same kinds of random perturbations and displaying behavior patterns that ought to be analyzable, and predictable to a degree, by any computer that was small, fast, and suitably programmed for the task.

Between 1976 and 1981, accordingly, Farmer and Packard developed computers tiny enough to fit into the heel of a shoe, and they wrote software that simulated the dynamics of a roulette ball closely enough to predict its exact landing spot a fair percentage of the time. But every scientific hypothesis had to be tested, and to submit this one to experimental trial, Doyne and Norman took themselves to the casinos of Las Vegas, Nevada, where, as it happened, they enjoyed a modest degree of success at the roulette tables.

All of this real-life physics experience would stand them in good stead later on in the early 1990s, when they too migrated to Santa Fe and opened a new business, the Prediction Company. The company would eventually operate out of a building that had once been a whorehouse. At the time they transformed it for higher (but not much higher) purposes, it was owned by Stu Kauffman. The corporate mission of the Prediction Company was a major extension of their roulette-wheel operation. This time, by the use of far more ambitious chaos-based scientific software, Doyne and Norman wanted to predict, outwit, and make a killing on the stock market.

Before acquiring the wherewithal for that venture, Farmer and Packard were forced to serve out their sentences by doing physics research of a more conventional sort. Packard went to the Institut des Hautes Etudes Scientifique outside Paris, where he did further work on chaos theory. Farmer, for his part, went to the Los Alamos National Laboratory, where he took a position with the Theoretical Division's Center for Nonlinear Studies and did further research on complex systems.

With this history behind them, it would have seemed that when Doyne Farmer came upon Stu Kauffman at a scientific conference in Bavaria, it should have been an encounter that went nowhere: king of the roulette wheel meets the grand master of self-organization, autocatalysis, and complexity. But in fact the two men were on precisely the same wavelength. Kauffman was fascinated by chaos, roulette, and even dripping water faucets. Farmer, similarly, couldn't hear enough about light-bulb networks, autocatalytic sets, self-organization, and the origins of life. The two of them spent a day hiking in the Alps, ascending into ever higher reaches of earthly atmosphere and scientific speculation.

Farmer invited Kauffman to Los Alamos to give a lecture or two at the Center for Nonlinear Studies. Kauffman showed up and talked a blue streak, as usual. Not long after that initial visit, Kauffman became a consultant to the Los Alamos National Laboratory.

DURING STU KAUFFMAN'S tenure as a consultant at the Los Alamos lab, he collaborated with Doyne Farmer and Norman Packard on a couple of projects. One was about the autocatalytic replication of polymers— long chainlike molecules made up of linked simple molecular building blocks called *monomers*. That work elaborated Kauffman's notion of simple molecular starting points bootstrapping themselves into more complex ones, which then reproduced themselves by autocatalytic means, which was his core explanation of how life arose. The other project was an equally speculative paper about viewing adaptive dynamic networks as models of the immune system. Both of these efforts were abstract in the extreme, examples of basic or "pure" science.

And then in a pattern that would recur time and again throughout his career as a scientist, Kauffman left the realm of abstraction and theory for the more prosaic world of disease and medicine. By the end of it, he would invent some new ways of making drugs and vaccines.

In this case, the transformation occurred on the island of Crete in the Mediterranean, off the coast of Greece. He'd gone there for a conference on fruit flies in the mid-1980s, but in truth he no longer found fruit flies the fascinating creatures that he once did. During a lull in the proceedings he wandered out into the parking lot, looked off toward the mythic wine-dark Aegean, and began thinking about the relationship between hormones and receptors. That relationship was fundamental to both disease and health: hormones were chemical messengers that coursed through the body, while receptors were nerve endings that could be stimulated by contact with a hormone that fit the receptor in the canonical lock-and-key fashion. An excess or deficiency of hormones was characteristic of certain medical conditions, the treatment for which sometimes required an external supply of the hormone in question. Estrogen replacement therapy, to treat the symptoms of menopause, was a classic example.

In the parking lot on Crete, Kauffman hit on a way of producing large quantities and varieties of artificial hormones, molecules that could work as well as and perhaps even better than the original hormone itself.

"The following things came to me," Kauffman recalled much later. "Suppose you have a hormone like estrogen, and suppose you have the estrogen receptor. Think of estrogen as a key and the receptor as a lock. Now throw away estrogen and ask, can I find the molecule that looks like estrogen? Well, let's make a hundred million random small proteins called peptides and ask whether any of them can bind to the estrogen receptor. If so, it's the second key that fits the same lock that estrogen does. So the keys must look alike. So I'm on my way to making a drug that could mimic or modulate, agonize or antagonize, the effect of estrogen. And I said to myself, *My God, I've just thought of an entirely new way of making drugs.*"

That was Kauffman's overall plan: generate lots of hormone-like molecules and then select from among them those that fit the receptor of interest. Conceivably, this was a way of inventing some powerful new cures for some age-old diseases. There was the minor detail of manufacturing "a hundred million random small proteins called peptides," but Kauffman had already figured out a way to do that.

The first step was to create millions of different DNA molecules, a task that itself might sound tricky or impossible. But as Kauffman well knew, there were already standard molecular-biology lab techniques, complicated but nonetheless entirely doable, for performing that miracle. The next step was to take those random DNA molecules and put them into bacterial cells, placing each separate gene sequence into a different cell. That too was no big deal, as there were likewise standard lab techniques for placing DNA sequences into cells.

Once inside the cells, the DNA sequences would proceed to play out their allotted roles in life, which was to be expressed in the form of whatever specific peptide, polypeptide, or protein the respective DNA sequence coded for. DNA was a recipe for making those molecules, and the cells were the tiny factories that did the molecular manufacturing work.

This gave Kauffman his desired millions of small proteins. Those millions in effect constituted an enormous warehouse of new proteins, a vast "library" of new molecules, some of which might prove to be medically useful.

Still, creating a library of peptides was only half the battle. The other half, and the greater challenge, was separating the wheat from the chaff, finding from among those initial millions the specific molecules that were of medical interest.

And finally there was the problem of actually doing all this in the lab, of effecting a proof of concept. Here, unfortunately, Stu Kauffman was somewhat at a loss. "I'm a really good theorist," he had said more than once, "but a grade B experimentalist."

But Kauffman had plenty of friends, colleagues, and connections and knew just the man for the task—Marc Ballivet, a molecular biologist at the University of Geneva, Switzerland. Unlike Kauffman, Ballivet was a lab whiz.

The two had met long before, in 1961, when Marc's sister, Christine, had brought Stu home to her parents' house in Nice after having met him in the Austrian Alps. This was back in Stu's salad days, between Dartmouth and Oxford, when he was living in his VW camper in the parking lot of the Hotel Post. Stu and Marc, it turned out, shared common interests in biology and women, and the two of them hit it off.

In the early 1980s, some twenty years later, Stu took a sabbatical leave from the University of Pennsylvania to work at Ballivet's lab in Geneva. Marc listened to Stu's scheme for creating libraries of peptides and proteins, and his scheme for making second keys to fit molecular locks, and realized at once that there was indeed considerable potential here for making new drugs, vaccines, and the like. As for selecting the right keys from the millions of new peptides, Marc Ballivet had an idea for how to do that, and he was more than willing to give it a try in actual lab glassware.

He knew he could make millions of random DNA sequences in vitro in a variety of ways, so long as he had sufficiently purified enzymes; none of that was any problem. Once he'd produced the DNA sequences, he could attach them to plasmids (tiny circular strands of DNA) and place the plasmids into bacterial cells. The cells would then go on to generate the various molecules coded for by the random genes. Thus far, all this involved standard lab techniques in molecular biology.

Ballivet's next trick, however, went beyond them. He decided that he would design the random DNA sequences in such a way that the resulting peptides would be expressed on the outside surfaces of the bacterial cells. The bacteria would grow the new proteins as little bumps on the skin, bumps that, if they were the right size and shape, would bind to the hormone of interest, estrogen, for example. Then he'd simply let the bacteria grow and reproduce themselves on a petri dish that he'd coated with estrogen.

The "good" and the "bad" protein molecules would then self-select themselves out in the following way. When sluiced with a liquid, the cells that had grown proteins of the "wrong" shapes (those with no affinities with estrogen) would be washed off the petri dish easily, since there would be nothing to hold them to the estrogen substrate. By contrast, the bacterial cells that had the "right" molecular bumps on their skins—bumps that fitted into the cracks and crevices of the estrogen molecules—those cells would cling to the estrogen substrate and be left behind in the petri dish.

Out of the millions of random peptides, therefore, only those that interacted with estrogen would remain.

Over a period of weeks, Kauffman and Ballivet experimented with these and other bacterial growth and screening techniques and discovered that some of them actually worked as planned.

They now decided that they had jointly invented a new technology, a practical method for synthesizing molecular keys to fit medically significant molecular locks. Since the technique could be used to discover new drugs and vaccines, the obvious next step was to patent it. Starting in the spring of 1985, therefore, Ballivet and Kauffman filed patent applications with the patent and trademark offices in several countries.

PART TWO

DATA

6

THE DATA DUMP

BY THE TIME the so-called Ballivet-Kauffman patents were issuing from the world's patent offices in the late 1980s and early 1990s, the world of science was fairly drowning in data. The latter years of the twentieth century had been characterized variously as the age of molecular biology, the age of the genome, and the information age, but whatever else it was, it was the age of data collection in science. In the dim past scientists confined their observations to the visible universe and recorded the observations in lab notebooks by hand. Such records, pictures drawn one by one or lines of text written one at a time, were easily intelligible to anyone who had the necessary background knowledge.

But instrumentation and computers soon changed all that. By the end of the twentieth century, scientists could detect elements of reality that had been inaccessible in all past ages: they could see the world's smallest structures with the aid of electron microscopes, atomic-force microscopes, tunneling microscopes, and other devices. And the range of observable reality far exceeded the visible, as scientists took observations in the infrared, ultraviolet, x-ray, gamma-ray, radio wave, and other regions of the electromagnetic spectrum.

As for storage, the lab notebook had been far outstripped by the computer. Indeed, by the beginning of the twenty-first century the tools of scientific observation had become so powerful, common, cheap, and easy to use, and the techniques of storing the resulting mother lodes of observational material had become so widespread and effortless—that is, you merely saved your latest mountain of data to a hard drive, compact disk, DVD, or some other form of auxiliary information storage—that scientific data was accumulating at a pace that far surpassed the human ability to understand, interpret, or make any practical sense of it.

This was the age of the data dump. There were DNA sequences by the billion from the Human Genome Project. There were astronomical facts, figures, and images by the gigabyte from the space shuttle, from the Hubble space telescope, from the International Space Station, the Voyager probes, the Venus flybys, the Mars landers, and increasing numbers of other off-planet, automated data collectors. Meanwhile, back on the home planet, amateur astronomers pointed their telescopes at the sky; downloaded their star data, planetary data, asteroid data, comet data, satellite data, and more into their PCs and their Macs; and let it sit there and gather dust, forever. Professional astronomers were even worse. The Sloan Digital Sky Survey, an attempt to map every visible object in the night sky, was expected to produce 40 terabytes (1 terabyte equals 10^{12} bytes) of observational data before it was finished. The Large Synoptic Survey, an even more ambitious project, was predicted to be producing 10 petabytes (1 petabyte equals 10^{15} bytes) of data per year, by 2008.

When it came to scientific data collection, however, particle physicists were by far the worst offenders of all. By the year 2000, a single particle research center, CERN, in Geneva, Switzerland, was already producing more than 1,000,000,000,000,000 bytes (10^{15}, or 1 petabyte) of numbers per annum. Soon the world's most powerful particle accelerator, the Large Hadron Collider at CERN, was turning out 100-petabyte data sets at a clip. The Library of Congress, by contrast, had a comparative scarcity of information housed on its few tiny miles of shelves, for the billions of words in all of its millions of books added up to a mere thousandth of a petabyte, or 1 terabyte's worth (1 trillion bytes) of data. That was nothing.

What all this meant in real-life terms was hard for the layperson to appreciate, for the units of measurement familiar to the average computer user had long since been bypassed by those coined to express the magnitudes of the more colossal data dumps. In the beginning were the lowly, basic, baby units, well known to every computer user: bits (binary digits), bytes (8 bits), and kilobytes (10^3, or 1,000 bytes). Then there were the two bigger step-ups in storage volumes that most computer users were familiar with: megabytes (10^6 bytes, average floppy-disk size) and gigabytes (10^9

bytes, the units of hard-drive storage space). Beyond that, however, you were in essentially alien territory, with terabytes, petabytes, exabytes, and god only knew what other wholly unrecognizable mega-power-user units, all of them having been defined well in advance by the world's official computing authorities.

The fact of the matter was, however, that the scientific data filed away in all these enormous volumes of computer storage space was, by itself, a totally useless commodity. The stuff just sat there on disk, hard drive, compact disk, or whatever, dead collections of ones and zeroes, on and off states that did absolutely nothing and were of no practical benefit to anyone or anything, whether human, animal, or plant.

To be usable, to be beneficial, indeed to be even remotely meaningful in any human sense, data had to be converted into something else: it had to be converted into *information*. The difference between data and information was the difference between viewing the diffraction-pattern color spectrum produced by light falling on the bumps of a compact disk, and playing the disk to hear the music its hills and valleys encoded.

Information in this sense—intelligible, deciphered, understandable data—was different from information in Claude Shannon's use of the term. In the 1940s, Claude Shannon, a scientist working at Bell Labs, had used a combination of mathematical theory and electrical engineering principles to show how information of any sort could be reduced to discrete impulses and transmitted reliably over telecommunication lines. Having defined and solved that problem, Shannon became known as "the father of information theory." But what Shannon meant by information was, essentially, *data*: uninterpreted strings of ones and zeroes, on states and off states, considered as such. The dots and dashes of Morse code regarded merely as dots and dashes, smoke signals taken as distinct and individual puffs of smoke separated by spaces of clear air—all these were equivalent phenomena, all of them were "information" in Shannon's sense of the term, regardless of the actual message, if any, that was conveyed by the series of blips. A string of random numbers was "information" for Shannon, even though it contained no actual concept. For him, the random-number series qualified as information so long as it was received precisely as sent.

The problem with information in Claude Shannon's sense was the problem of faithfully separating signal from noise, differentiating the genuine dots and dashes from the background static and distinguishing the true puffs of smoke from random wisps of fog or from banks of low-hanging clouds. At the turn of the twenty-first century, the "problem of information" in that sense had long since been solved in principle, largely by Shannon himself. It was essentially a matter of ensuring accurate data transmission through the use of redundancy, filtering systems, pattern-detecting algorithms, and so on.

Still, some of the data banks that scientists had accumulated by the turn of the twenty-first century were so gigantic that even applying the pattern-detecting algorithms to them was no small matter. In some cases, the volume of data on hand vastly exceeded the computing power available to the scientists who had collected it, and in consequence the researchers involved were prompted to invent some imaginative techniques for interpreting their own observations.

There was the SETI@home project, for example. SETI, the Search for Extraterrestrial Intelligence, went back to 1959 when astronomer Frank Drake began a search of the skies, looking for intelligent radio signals from outer space. Intelligent aliens, he theorized, would be trying to communicate with us, or with others of their kin, by broadcasting radio signals that would be easy to distinguish from natural sources of radio emission as well as from ordinary cosmic background noise and static.

Drake started his project at the National Radio Astronomy Observatory's twenty-six-meter antenna at Green Bank, West Virginia. He pointed the antenna at the two closest solar-type stars, Tau Ceti and Epsilon Eridani, and listened for signals. He listened for several hundred hours, during which time he experienced a couple of hair-raising false alarms, thinking he was hearing messages from an intelligent culture beyond the earth. In the end, all the false signals proved to be of terrestrial origin.

There had been other such programs in the years since, both public and private, with a similar lack of positive results. Then in the late 1990s, a group at the University of California, Berkeley, began Project SERENDIP—the Search for Extraterrestrial Radio Emissions from Nearby Developed Intelligent Populations. The group would use the large radio telescope at Arecibo, Puerto Rico, to make a systematic search of the Northern Hemisphere's skies, looking for alien radio transmissions. The scientists stored the results of the sky survey on 35-gigabyte digital tapes, each of which held about fifteen hours of incoming data. The entire SETI sky survey would require eleven hundred tapes, making for a grand total of 39 terabytes of recorded material, a volume equivalent to thirty-nine Libraries of Congress.

The question was, how did you analyze 39 terabytes' worth of data looking for what might turn out to be a lone cry from an alien culture when you lacked the necessary computing power? The answer was by distributing the data to millions of computer users who collectively would provide the needed computer-processing capacity. The plan was to break up the data into manageable chunks and have volunteer home computer users process those chunks during computer downtime by employing an algorithm that worked as part of a screensaver utility. Both the raw search data and the screen saver would be supplied to volunteer home users over the Internet free of charge.

The result was Project SETI@home, funded largely by equipment donations from Sun Microsystems (who supplied the servers) and Fujifilm (who supplied the 35-gigabyte digital tapes) and by gifts from Microsoft cofounder Paul Allen and Intel cofounder Gordon Moore, author of "Moore's law" (the claim that microchip processing power doubled every eighteen months).

As a plan, it was crazy enough. And indeed, it succeeded beyond the wildest expectations of the project leaders. By the end of the year 2000, SETI@home had acquired 2,438,045 volunteers who altogether had spent a combined total of 437,000 years of computing time analyzing the search data and looking for pulsed signals.

The volunteers found, if anything, far too many of them—some 1.1 billion candidate signals. The home-based SETI workers sent their candidate signals back over the Internet to the SETI@home research staff, who analyzed them with additional algorithms in an attempt to discover which, if any, were of intelligent origin. No signal passed all the tests, however, and the SETI@home Web page continuously posted the statement: "As of yet, SETI@home has not detected any radio signals that indicate the presence of extraterrestrial intelligence."

Known as *distributed computing*, the SETI@home data-processing technique was later adapted by other scientific groups including the National Foundation for Cancer Research, which was hunting for a cancer cure, and Oxford University, whose researchers were looking for a drug that would neutralize anthrax toxins.

Successful as it was, such publicly distributed data processing was not an acceptable technique to all scientists. In particular, scientists employed by private, profit-making firms such as drug companies, agribusinesses, and chemical firms did not have the option of sending their raw data out over the Internet for all and sundry to process in their spare time at home. To such researchers, data equalled dollars and could not be broadcast indiscriminately to anyone willing to process a portion of it. Somewhere in those data caches, conceivably, lurked the next miracle drug, a major new source of revenue.

Clearly, there was some money to be made here, for corporations interested in locating buried treasure were willing to pay virtually any amount of ransom to find it. The Info Mesa arose in part to satisfy that demand for interpreted, mined, processed data, data that had been converted into information, into actual knowledge.

ALTHOUGH HE HAD invented the SMILES language specifically in order to manage the burgeoning chemical data sets required by the EPA's Toxic Substance Control Act, Dave Weininger himself, ironically, would never use the SMILES system for that purpose. No sooner had he invented it in 1981 than the incoming Reagan administration, which placed its priorities elsewhere, derailed much of the grand environmentalist mission. In 1982, as a consequence, Weininger left the EPA and took a job with the Medicinal Chemistry Project at Pomona College in California.

Pomona's Medicinal Chemistry Project was an effort that went back to 1964, when Corwin Hansch, a Pomona chemist, and Toshio Fujita, of Kyoto University, figured out a method for predicting the reaction rates and equilibrium conditions for a variety of reactions among organic chemicals. Both of these items were of great practical utility to working chemists but were especially so to drug developers, who needed to know the rates at which a candidate drug would dissolve in water or other solvents and then be absorbed by the digestive system and taken up by the target cells. Chemical reactions took place at rates that varied with the nature of the reactants, their temperature, and many other factors, but precise information on such parameters was often hard to come by.

Hansch and Fujita had the idea of collecting data on chemical reaction rates and providing it to chemists for a fee.

And they would do the same thing, they decided, with data on chemical equilibrium conditions. In chemical reactions, a state of equilibrium was reached when the reactions in question balanced each other so that there was no further net change among the reactants. In the simplest case, the respective amounts of gas and liquid in a closed vessel reached a state of equilibrium when the rate at which the liquid molecules escaped into the gas equaled the rate at which the gas molecules dissolved into the liquid. As with information on reaction rates, chemical equilibrium data was bread and butter to drug designers, as well as to manufacturers of anesthetics, pesticides, fertilizers, and many other types of bioactive chemicals, but as yet no small, practical, self-contained listing of it was available in any form.

During the mid-1960s, Corwin Hansch, together with a postdoctoral researcher at Pomona by the name of Al Leo, began the Medicinal Chemistry Project—MedChem for short—gathering data on reaction rates, chemical equilibrium conditions, and a few other crucial parameters and offering it to biological and medicinal chemists who could use it in their search for new drugs. They entered the information into a database in the form of Wiswesser Line Notation, and made the resulting files available to subscribers in any of three formats: hard-copy computer printout, nine-track magnetic tape (from which users could make a hard copy), and microfiche.

The MedChem subscription service was highly successful, and the project leaders used the proceeds to gather data on a further important consideration to drug designers, a measurement called the *partition coefficient*. For a drug to be biologically useful, it had to be transported to the "active site," the place in the body where it was needed, and in order to arrive there, the drug compound had to avoid getting stuck where it wasn't required and could do no work. If, for example, a candidate drug compound were excessively hydrophilic, or water-loving, it would tend to remain behind in the aqueous, or watery, compartments of the body, such as the stomach and the bloodstream, and never get to wherever else it was supposed to go. If, on the other hand, the compound were too lipophilic, or fat-loving, it would tend to collect in the body's fatty tissues, where little biological action of consequence would occur.

An ideal drug had to strike a balance between being hydrophilic and hydrophobic so that it could be transported to the active site by an essentially random-walk process without getting stopped at extraneous places anywhere along the way. The so-called partition coefficient merely expressed the degree to which a candidate drug compound was happy in both watery and nonwatery media, and as such, it was an important factor controlling the distribution of chemicals both in the body and in the external environment.

Unfortunately, the partition coefficients of the world's chemical compounds were only poorly known by chemists. It was possible for a chemist to determine the coefficients experimentally by placing the substances in solvents, shaking the flask, and then measuring the resulting concentrations, but taking precise measurements of the sometimes extremely dilute mixtures was time-consuming and often difficult. Chemists had already measured the partition coefficients for a number of important compounds, but those substances barely scratched the surface of the chemical compounds that were theoretically possible. Basically, there was no end to the types of substances that chemists could synthesize, and there were potentially millions of different compounds that could be investigated as drug candidates. As for the partition coefficients of the compounds that had not yet been created, there was obviously no way of measuring them in the lab, which meant that the only way to get them was by calculation. Al Leo had developed a method of calculating partition coefficients from the known values of a given compound's chemical constituents, but the procedure was slow, repetitive, and tedious.

In 1982, Al Leo met Dave Weininger at an EPA conference in San Antonio, Texas, and told him about his manual methods of calculating partition coefficients. Weininger, for his part, told Leo about his brand new chemical language, SMILES. Being on intimate terms with both chemistry and computers, Dave now decided that he could write a program that could calculate partition coefficients automatically for any substance for which adequate structural information existed. Such a program, both he and Al Leo realized, would be a bonanza to the ranks of medicinal chemists.

Weininger moved to Pomona and over the course of the next six months proceeded to download from Leo everything he knew about partition coefficients, hydrophobicity, the molecular electrical charges that give rise to hydrophilic behavior among molecules, and so on. In the end Weininger produced an expert system that embodied this knowledge and that was able to calculate partition coefficient values for virtually any substance in an initial database of approximately four thousand compounds. You simply typed in the SMILES for the compound in question, let the program run, and in a matter of seconds, out rolled the answer. For a novel substance, or one not in the database, the program automatically analyzed the molecule, looked up the values for each known fragment, and then reassembled the fragments, making sure to account for any chemical interactions that might take place between them.

Weininger called the program *ClogP* (pronounced "C-log-P"), standing for "*c*alculated *log*arithm of the *p*artition coefficient." (When expressed as a logarithm, a partition coefficient was known as "logP.") When it was finally debugged, ClogP worked like a charm—except for the "missing fragment" message that unfortunately cropped up every so often. In such cases, the program merely estimated the fragment's partition coefficient and then incorporated it into the main calculation.

For people who made their living as medicinal chemists, ClogP was such a useful tool that few wanted to be without it, even despite the occasional "missing fragment" report. Pomona's MedChem project offered the software as a commercial package, and ClogP was soon adopted by the EPA, Dow Chemical, and eventually many of the rest of the world's pharmaceutical, agricultural, and chemical companies. (In 2001, ClogP Version 4.0, now advertised as having "no missing fragments," was priced at $2,500 for commercial users, $1,875 for governments, and $1,500 for academic institutions.) It was Dave Weininger's first big commercial success at reducing the chemical universe to information, and chemistry to an information science. Many others were soon to follow.

In 1984, since he was now making a regular salary at Pomona, Dave bought his first private plane, a 1966 Alon A-2 Aircoupe. This small, single-engine, low-wing aircraft seated two people side by side. One of the plane's distinguishing features was a sliding canopy: you slid it back and stepped over the side to get in, which was unusual for a private plane—most small aircraft had conventional automobile-style door-entry systems. The sliding canopy made the Aircoupe seem more like a jet, or at least a high-performance aircraft. Plus, whenever he wanted to feel like an old-time barnstormer, Dave could slide the canopy back in flight, exposing him to the sky, the wind, and the roar of the engine.

Dave loved the craft, which he named *Puer Æternus*. He ended up fly-ing it all over California and the nearby states, and once even to Quebec, which he remembered ever afterward as "a long damn flight." The little craft was his one major extravagance to date.

A couple of years after moving to Pomona, Weininger met a computer engineer by the name of Yosef Taitz, known to one and all as Yosi. Yosi had been born in Riga, the capital of Latvia, on the east coast of the Baltic Sea. His family had later immigrated to Israel, where Yosi ended up liv-ing for fourteen years. Yosi was extremely serious about his ethnic iden-tity and refused to own or even drive vehicles made in Germany. At length he joined the Israeli Air Force, which is where he started learning about computers.

Yosi came to California in 1980 and got a degree in computer engi-neering from California State Polytechnic University in Pomona. He worked as a computer engineer for a while, for a company that built peripherals. He soon decided that he would rather be working for himself, and so he started studying for an M.B.A. at the University of California, Irvine. At about the same time, he founded a consulting business called Computer Systems Technology, in Mission Viejo, California, just south of Los Angeles. The company sold UNIX-compatible peripherals for Sun Microsystems computers. One of Yosi's clients was the Claremont Col-leges, a group of seven colleges, including Pomona College, home of Dave Weininger and the MedChem project.

By this time, 1985 or so, Dave had been anointed with the vision that would drive him for the next several years, the prospect of doing chem-istry not in the lab but in the computer—a sort of virtual chemistry, or as he sometimes thought of it, "in silico" chemistry. His ultimate fantasy was that a chemist would be able to see, on the screen, representations of vivid, color-coded, three-dimensional molecules, objects that you could move around with the mouse, objects that you could even bump together, like on-screen billiard balls, whereupon, in a flash and almost as fast as it happened in the real world, the computer would calculate and then graph-ically show you what would happen and what the chemical output would be, the molecular result of the chemical reactions that would have taken place.

This idea was not exactly new: the field of "computational chemistry" had been around for a while, but so far it had never been implemented in quite this extremely sophisticated and futuristic fashion. The molecular modeling programs that existed at the time did not enjoy wide acceptance among working chemists who regarded their predictions as often wrong, or at the very least as unreliable and imprecise. They were the kind of thing you had to take with a huge grain of salt even when you were dealing with small, well-known molecules, much less with larger, unconventional, or totally unknown structures. For this reason, some chemists dismissed such programs as "Nintendo chemistry."

Dave Weininger, however, wanted to go far beyond all that. He wanted to create a program, or a suite of programs, that rivaled the precision of experimental chemistry as practiced in the lab. Key to it all would be facts, data, *information*. Dave planned on cramming a powerful computer with the sum total of known scientific facts and figures about the world's chemical substances, and programming the machine to make proper sense of it all.

He would proceed in stages, starting with chemical databases, look-up tables, and programs that reproduced simple and basic chemical processes. ClogP was a case in point, but that was merely the beginning. His next step would be to store in a computer's memory—in RAM—all of the chemical parameters that characterized a wide range of molecules. For this he needed a computer with lots of memory, far more than was available in the standard personal computers of the day. Weininger went over to the college's computer center, described his problem, and was told to see Yosi Taitz.

Dave and Yosi got along from the start. Both of them were calm, relaxed, no-stress types (without fear of contradiction, Yosi could even be called jolly), and they found it exceptionally easy to get along with each other. Even so, both individuals were highly intelligent, ambitious, perfectionistic, and driven.

Plus they had much else in common: Dave's brother was then living in Jerusalem, and Yosi loved airplanes and flying. Of course, it was inevitable that Dave would take Yosi up in the Aircoupe. Dave flew it to Fullerton Airport, halfway to where Yosi lived, and Yosi met him there for his inaugural flight in *Puer Æternus*.

"This was the first time that I touched the controls of a plane," Yosi recalled much later. "We went up into the clouds. Dave pulled back the canopy, I stuck my hand out and I touched the clouds. All of this was very memorable for me."

The two of them were also on the same wavelength as far as computers were concerned, and their talents complemented each other's: Yosi was the hardware man; Dave was the software genius. To provide maximum utility for the user of his computational chemistry systems, Dave wanted to implement his various programs in a truly portable computer, one that would fit inside a briefcase. So Yosi Taitz built him one.

"One of the companies I was consulting for at the time was National Semiconductor," Yosi remembered. "They had a plan for putting computers together by using stackable boards: one board is the CPU, the next is memory, the third a communication device, and so on. There is no common bus, you just plug one board into the other. So I put a computer together for Dave and put it into a 'Zero' case, a strong aluminum carrying case. It was transportable but it didn't have batteries, so it was not truly portable in the modern sense."

It also wasn't equipped with an excess of memory, but Dave nonetheless installed a few of his early programs—SMILES, ClogP, and some others he was working on—and they worked well enough. He demonstrated the system at a meeting of the MedChem User's Group in Claremont— and everybody who saw it was wildly enthusiastic and supportive: "Fantastic stuff!" "Clearly the future of chemistry!" "Got to have one!" And so on.

Whereupon Dave and Yosi decided to go into business together and offer the world a turnkey system for chemical computation. Yosi would do the hardware; Dave, the software.

"So we put together a prototype and we offered it to friends, associates, people in the pharmaceutical industry, former students of Corwin Hansch, members of the MedChem User's Group, everyone we could think of," said Yosi Taitz.

"And nobody bought it."

FOR DAVE AND Yosi, the miserable sales record of their prototype turnkey chemical information system was a learning experience. Lesson one was that two rather suspect individuals such as themselves did not make, from a sales standpoint, an entirely plausible corporate image. Here was Dave Weininger, habitually dressed in amorphous ripped sweatpants and faded T-shirt or, at his very most formal, in a pair of three-year-old jeans and long-sleeve sweatshirt. With his fuzzy-scale black beard and extremely long ponytail wrapped in a rubber band, plus motorcycle, he looked like a nightmare straight out of Hell's Angels. Yosi, for his part, although neatly frocked, clean shaven, and more fastidious looking in every way, was nevertheless a foreigner who spoke with a combination Latvian–Israeli–southern Californian accent, and no part of the overall picture that the two of them jointly presented was exactly awe inspiring to potential customers. Further, Dave kept strange hours and worked out of his house in Pomona; Yosi worked more or less nine-to-five but operated out of his own home in Irvine. Besides, all they had to show for their efforts was that one lonely prototype system. No vast production line stood behind them, no factories, no proven hardware, no software, no string of smiling and satisfied customers. So, in all honesty, who could blame people for not buying their stuff, which in demos seemed to work like some sort of chemical black magic? It was all just slightly too slick to be believable.

The other take-away message was that what they should really concentrate on was software. Corporations had their own computers that they had already paid good money for and placed lots of confidence in. There was no point in their trying to compete with a company's own precious and trusted machines. Dave and Yosi would simply fill those machines with their own proprietary chemical information systems, for a fee. That's what their target customers would be interested in, anyway—chemical information, not more hardware.

And to give their whole package maximum credibility, they decided that they had better form an actual business structure in the standard fashion. So they met one night in a restaurant and, in the finest Hollywood tradition, sketched out their business plans on paper napkins. First of all, each would make his own private list as to why they should work together and start a company, and then they would compare lists.

Dave wrote down his first reason: "To have fun."

Yosi wrote down his first reason: "To have fun."

Dave's second reason was "to change the way chemistry is done today."

Yosi's was "to run a successful company."

Neither remembered what the third reason was, but by the time they were finished with their lists it was clear that each of them was highly simpatico with the other, and that they should go ahead and form a company. They then divided up responsibilities: Dave would do software, Yosi would manage the business end, and together they would change the face of chemistry and perhaps even make a pile of money in the process. At all events they'd have fun!

In 1986, therefore, they founded Daylight Chemical Information Systems as a partnership. "Daylight" was an acronym that stood for Dave and Yosi and the divine *light* that would soon be emanating from their vast array of superlative and heavenly software. Their first product release would be called *Day One*; their second, *Day Two*; and so on. The company image would be the sunrise, perhaps symbolizing the dawn of creation. Having lots of fun already, they were nothing if not cute.

Yosi rented an office in Claremont, close to the college. It consisted of a large main room flanked by two smaller rooms on the second floor of an office building. They put folding tables around the perimeter of the big room, covered the tables with computer monitors, and regarded the result as Daylight intergalactic headquarters. To solidify matters businesswise, Yosi finished up the M.B.A. that he'd started long ago at the University of California, Irvine, finally getting it from West Coast University, a private evening college with a campus in Santa Ana.

Dave, meanwhile, perfected his latest product offerings. So far, he had created SMILES, his chemical nomenclature system, and ClogP, his program for calculating partition coefficients. But these were just preludes, mere overtures, to his grand synthesis of chemical information databases and computer-based chemical reaction processing. His next two software productions, Thor and Merlin, would be socko applications in every sense of the word.

Thor was the name of a figure in Scandinavian mythology, the god of thunder, rain, and farming. Thor was also the root name of thorium, a radioactive element, number 90 of the periodic table, that Berzelius had discovered in 1828. What Thor more properly stood for in Dave Weininger's master plan, however, was *Thesaurus Oriented Retrieval*, a module that would function as his main chemical database and storehouse of associated information. A chemist who wanted to know the structure of a given molecule or gain some insight as to its chemical behavior, or who wanted to see a list of scientific articles written about it, would be able to look up the compound in Thor, provided that the user had some sort of unique identifier of the item in question: its SMILES, for example, or its common name, such as "caffeine." The user would type in the identifier, press <ENTER>, and receive a complete and formal dossier on the chemical entity in question, including a structural diagram of the molecule as well as all sorts of numerical values of interest to chemists such as its partition coefficient, or logP.

The virtue of Thor was that it was able to look up the entity from an extremely large database of very complicated molecules—the Thor data set would eventually exceed 10 million separate chemical structures—and provide an answer in an incredibly short span of time, often in little more than a second. Dave accomplished this miracle by using an information storage and retrieval method known as *hashing*.

Hashing, so named because it seemed to reduce orderly data to hash, was a way of transforming long lists to smaller and more manageable arrays without the loss of any information in the process. Suppose, for example, that you had a list of five hundred items, each of which had a relatively long and complicated name, such as

hexamethylenediamine

amylopectin

b-keratin

zeaxanthin

bis(2-chloroethyl)thioether

sodium para-dodecylbenzenesulfonate

2,2,4-trimethylpentane.

Now if you wanted to retrieve a piece of information associated with a specific item on that list, you would have to run through the entire list, looking for the item by name, a time-consuming chore even for a computer. The process would be speeded up by ordering the original list according to a rule—listing the compounds alphabetically, for example. But that was not *hashing*, that was only *sorting*, and its effectiveness was limited by the fact that the sorted list was still almost as complicated as the original, unsorted list, and what was worse, the alphabetized list didn't easily accommodate items whose names began with a nonstandard character (such as a Greek letter, as in β-keratin) or a number (such as in 2,2,4-trimethylpentane), or other oddities so near and dear to the hearts of chemists.

Instead, hashing worked by assigning to each compound a number arrived at by applying a given mathematical function to each entry, and then making an ordered list of the numbers. For example, the function might convert each of the word's characters to its ASCII value, take the sum of those values, and then divide the sum by a large number. Listing the result numerically might produce, for example, the following table:

152	β-keratin
274	*bis*(2-chloroethyl)thioether
392	hexamethylenediamine
501	amylopectin
757	sodium *para*-dodecylbenzenesulfonate
844	zeaxanthin
979	2,2,4-trimethylpentane

Such an ordered numerical list would enable the user (or the computer) to locate specific items within it instantly, without having to search through the entire array word by word. The user would simply type in the name, press <ENTER>, and the computer would invisibly do the rest.

Unfortunately, almost any hashing algorithm yielded duplications of numbers—these were called "collisions" (or "hash clash")—but there were known ways of avoiding collisions, and Dave Weininger exploited them to the fullest. By adroit use of hashing algorithms and collision-avoidance techniques, Weininger was able to store in computer memory a far greater volume of chemical substances than would otherwise have been possible, making for exceptionally short information-retrieval times even for extremely large databases.

Still, impressive as it was when the software was finally up and running, Thor was nothing when compared with Merlin. Named after the magician of Arthurian legend, Merlin would be Daylight's search engine, and it would indeed be magical. Whereas Thor operated by looking up one structure at a time, Merlin worked with entire sets of structures. You could ask it, for example, to search an extensive data set and locate all compounds that were structurally similar to a given compound, or even to a known *fragment* of a compound. You could have it look for substances that reacted with a given compound in a specific fashion. You could have it look for textual data items that pertained to a range of compounds with similar chemical characteristics.

Merlin also differed from Thor inasmuch as you didn't have to know a given chemical's name, formula, SMILES, or any other identifier in order to use it. Instead, Merlin allowed you to search through the Thor database by starting with a picture. In a tour de force of programming, Dave created a GRINS application (*Graphic Input of SMILES*), by which the user could submit queries graphically. The user would simply draw a structure using keyboard and mouse, whereupon GRINS would come back with the chemical's correct SMILES, which it would then automatically submit for a Merlin search. In other words, the user would configure Merlin for the kind of search desired, draw a diagram using GRINS, press <ENTER>, and immediately get the information requested.

In the end, Dave got Merlin working so fast that it was able to search through more than 500,000 different chemical structures per second, looking for structural similarities, reactants, reaction products, mixtures, substructures, and whatever else the user might have a need for. The output could be either lists of structures, textual information, or graphics, with molecular fragments appearing in any combination of user-selected colors. A typical Merlin output might look like this:

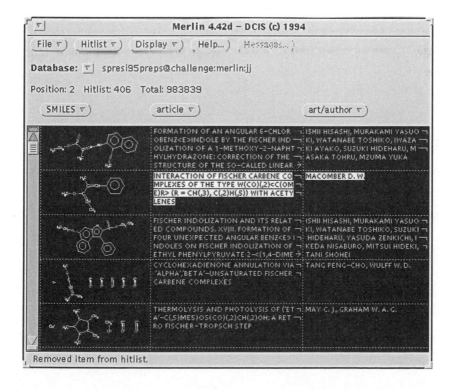

Before, when a chemist's primary lookup tool was the fifty or more volumes of *Beilstein*, each volume of which was more than five hundred pages long, such searches were not merely difficult—they were impossible. With Merlin, they were trivial. As a way of navigating through the data dump of modern chemistry, Merlin seemed like an occult application, one that entertained, baffled, and often impressed potential customers—of which there were finally beginning to be a few.

In 1986 Daylight was approached by a British company, V. G. Analytical, maker of mass spectrometers, which was looking for a way to differentiate itself and its machines from those of other mass-spectrometer manufacturers. By this time, word of mouth had gotten around about SMILES, ClogP, Thor, Merlin, and some of Dave's newer creations, and V. G. Analytical's technical staff was thinking about installing Daylight's software package (then known as Daylight Menus) on its machines so that they could display the structure of a molecule as soon as they got a hit. The machine would identify the molecule spectrographically, and the SMILES software would match the molecule to the SMILES table and then immediately generate the correct molecular depiction. No other mass spectrometer in the world would be able to perform such an amazing feat.

In negotiations with the British company, Yosi, the corporate sales expert, started off a bit on the high end, asking for an amount in the several-million-dollar range. This would have pleased Dave, since by that time he felt like he had written several million lines of computer code. Four months later, Yosi finally closed the deal. In the end, he sold the package for far less than what he had asked for, but that was the way things worked in the business world, especially in the startup stage.

Still, it was the company's first big sale, and it was clear that the sun had finally risen on Dave and Yosi.

7

MR. STARTUP

WHEN STU KAUFFMAN envisioned his system of creating novel proteins as a method of discovering new drugs, he was entering a field that stood in dire need of just such a technological advance, for at that time drug discovery was a black art in which the layperson's untutored view of the process was all too close to reality.

Pharmaceutical companies, according to the popular view, discovered their new compounds in roughly the following way: In his declining years, the company's chief executive officer takes an ecotour vacation in the Amazon rain forest. Deep in the swamps, he hears tales of a group of aborigines that since the beginning of recorded time have used an exceptionally potent snake-venom derivative to cure the blinding headaches that always followed their tribal fertility rites. The CEO bribes a local bush pilot to fly to the scene and collect a few samples of the stuff, which is then forwarded to the drug company's research headquarters by means of diplomatic pouch. Several decades later, after interminable delays caused by overzealous bureaucrats in the Food and Drug Administration (FDA), Imitrex finally hits the shelves—the last word in migraine relief.

That was the comic-book view of drug discovery, perhaps, but even at the end of the twentieth century it was not too far from the truth. Drugs were still discovered by accident, through folklore, old wive's tales, and stories of miracle cures wrought by various tree-bark extracts, just as they had been throughout recorded history.

Aspirin, one of the most commonly used of all drugs, went back to antiquity and the writings of Hippocrates, who told of a bitter powder derived from the bark of the willow tree that was effective in treating aches and pains. In fact, willow bark did indeed contain salicylate compounds, the ingredients of aspirin. In 1899, the Bayer pharmaceutical company in Germany had learned how to synthesize acetylsalicylic acid $(C_9H_8O_4)$, to which they gave the trademark name aspirin, and which they proceeded to manufacture in industrial quantities, thereby making a modern commercial success out of an ancient folk remedy.

In the 1630s, tales reached Europe of a miraculous "fever tree," found on the slopes of certain mountains in Peru. An extract from the tree's bark, according to the legend, was supposed to cure the "ague," the old word for malaria. The plant turned out to be the cinchona tree, known in Peru as *quina quina*. The bark of the *quina quina* tree would prove to be the source of quinine, the first successful antimalarial drug.

And then in 1775 a British physician with an interest in botany, William Withering, heard a literal wife's tale of how the practice of ingesting the dried leaves of foxglove, a large flowering plant indigenous to Europe, was successful in treating cases of dropsy, an accumulation of bodily fluids caused by failure of the heart to pump sufficient amounts of blood through the circulatory system. Such a use of the foxglove plant, he said, "had long been kept a secret by an old woman in Shropshire, who had sometimes made cures after the more regular practitioners had failed."

Withering experimented with the substance himself, found that it worked precisely as claimed by his source, and in 1785 wrote a treatise called *An Account of the Foxglove and Some of its Medical Uses: with Practical Remarks on Dropsy, and Other Diseases*. He had discovered the drug digitalis, another triumph of herbal folklore.

Penicillin was by far the most celebrated example of accidental drug discovery, however. In 1928, during an attempt to identify the microbe responsible for the 1918 flu pandemic that had killed 22 million people around the world, Alexander Fleming, a Scottish physician and bacteriologist, had grown some staphylococcal bacteria cultures at the University of London hospital. He left the cultures standing in open petri dishes for several days while he went away on vacation. According to the standard account, the petri dishes had been left unprotected, near an open window.

Upon his return to work Fleming was about to get rid of the samples when he noticed that some mold had fallen onto the petri dishes, contaminating the cultures. That was only to be expected, given the state in which he'd left them. What was unusual about it was that there were dark spots around some of the bacterial colonies—dead zones, as if the mold that had landed on the cultures had killed off the bacteria. Moreover, no new organisms of any kind had arisen on the dead spots in the interim.

Fleming isolated the substance responsible and identified it as the fungus *Penicillium notatum*, standard-issue bread mold. The fungus apparently secreted something that inhibited bacterial growth. He didn't know what it was chemically, but he named the active substance "penicillin," after the mold. He cultured more of the stuff and found that the penicillin continued to kill staphylococcal bacteria even after being diluted as many as eight hundred times. Not only that, it did no harm to human white blood cells at any concentration that was lethal to the bacteria. Penicillin, in other words, seemed to be an exceptionally selective magic bullet that killed foreign invaders while leaving the body's healthy cells intact.

Mass production of penicillin began in the 1940s, whereupon once deadly diseases, such as syphilis, instantly became curable, and the compound entered the vernacular as a miracle drug. Penicillin soon became one of the most important pharmaceuticals in the entire practice of medicine—all from an accidental deposit of mold that wafted in through an open window.

The most astonishing part about all this ancient history, however, was that even in the modern era, right into the 1990s and into the twenty-first century, new drugs were still being discovered by the same old time-honored methods of accident, serendipity, and trekking through jungle in search of secret medicinal herbs. In 1992, one such herb became a candidate drug against the AIDS virus.

In the early 1970s, Paul Alan Cox, a Mormon in his twenties, was doing missionary work in Samoa. "Our duty was to explain the beliefs of the Church of Jesus Christ of Latter-day Saints to any who were interested, and to assist the people in humanitarian causes," he said later. "It was a wonderful time of spiritual growth for me."

After a two-year period in the tropics, Cox returned to the United States and entered a graduate program at Harvard, from which he received a doctorate in botany. Later he accepted a teaching position at Brigham Young University in Provo, Utah. Finally, in the late 1980s, he returned to the South Pacific, visiting Polynesia and American Samoa, but this time

with a new goal in mind. Many of the world's drugs, Cox had learned during the course of his studies, had originally come from plants. "Plants have
been a rich source of medicines because they produce a host of bioactive
molecules, most of which probably evolved as chemical defenses against
predation or infection," he said. Indeed, Cox once estimated that during
the 1950s, approximately 120 commercially sold preparations, accounting
for some 25 percent of all prescriptions filled in North America, were for
drugs that had originated in the plant world. He also estimated that out of
the earth's estimated 250,000 flowering plant species, fewer than 1 percent had been studied for their possible healing effects. When he went
back to Samoa, therefore, it was for the purpose of examining the local
plant life and talking with the indigenous healers in a search for potential
new drugs.

By 1984 he thought he might have found one. On the island of Savai'i
in Western Samoa, several local healers told him about a medicinal brew
made from the wood of a rain-forest tree called *Homalanthus nutans*.
Supposedly, an extract from the tree cured hepatitis and yellow fever,
among other things.

That claim stopped Cox in his tracks: both hepatitis and yellow fever
were caused by viruses, and even in the late twentieth century, Western
medicine boasted very few drugs that were effective against viral diseases.
Immunoglobulins and vaccines were successful against some types of hepatitis but had no effect on others, for which the current medical wisdom
had it that "the only treatment is prevention." As for yellow fever, that disease could be prevented by vaccine or controlled by eradicating the mos-
quitoes that carried the virus, but once a person had contracted yellow
fever, no medication was effective against it.

Except, perhaps, for this exotic folk preparation made from the *Homalanthus* tree.

Cox discovered that *Homalanthus nutans* was a relatively small tree that
usually grew to a height of ten feet or so and flourished along the edges of
the Falealupo Rain Forest. The tree had spade-shaped leaves with a
whitish sheen on the bottom. The leaves were attached to the branches
with exceptionally thin stems, giving them the appearance of being suspended in air.

According to the "taulasea," the Samoan herbal specialists, the *Homalanthus* leaves were used in water infusions to relieve back pain and

abdominal swelling; the roots, to suppress diarrhea; and the stem wood, to treat yellow fever. To learn more about the latter, the taulasea advised Cox to see an elderly healer by the name of Epenesa Mauigoa.

"I know a little," she said, when Cox asked if she knew anything about the yellow fever medicine. In fact she was a walking herbarium of Samoan remedies and ended up giving him 121 recipes that drew on more than ninety different plant species. The one for the yellow fever potion was merely number 37 on her list.

Start with a trunk of the tree, she told him, and then "macerate it into a cloth and plunge the cloth [like a tea bag] into boiling water. Decant the infusion into a cup. Sweets and greasy things are forbidden to the patient. Also forbidden is anything with coconut fat."

Well, why not? Cox decided to bring samples of *Homalanthus* wood back to the United States. It was customary in ethnobotanical research to preserve wood and leaf samples inside herbarium vouchers, essentially two boards pressed together, a technique that hadn't changed much since the era of the earliest explorers to the New World. So Cox now took branches and leaves from the *Homalanthus* plant, flattened them between sheets of newspaper, placed the sheets between felt blotters, separated the blotters by sheets of corrugated cardboard, and inserted the whole sandwich into his plant press, two large boards bound with strong adjustable straps. To get rid of excess moisture that might damage the samples, he dried his plant press over the fumes of a kerosene stove.

He also prepared a sample of chopped-up bark and stem wood, packed the pieces into a Sigg bottle—a lightweight aluminum canister used to store everything from drinking water to gasoline—and filled the bottle with 75 percent ethanol. Then, with a solvent-proof marking pen he wrote on the outside of the bottle "#842 Homalanthus, stemwood."

Back in the United States with his samples, Cox tried for a while to interest some of the major drug companies in evaluating the stuff but had no luck. "Because of the prevailing attitudes at the time," he said, "no drug companies were inclined to analyze the specimens." Despite the success of a few drugs that had emanated from an oral tradition, pharmaceutical companies took a dim view of unidentified substances coming from tribal shamans living on islands lost in the Pacific—and who could blame them?

Cox sent his samples to a friend of his at the National Cancer Institute (NCI) in Bethesda, Gordon Cragg, who had agreed to test them for possible anticancer activity as well as for effectiveness against the AIDS virus, HIV-1.

In 1992, eight years after he had first come upon the plant in Samoa, Paul Alan Cox received a letter from Gordon Cragg telling him that the concoction was extraordinarily active against HIV-1, at least in the test tube. Cragg and his colleagues then isolated the plant's active component, a chemical compound called *prostratin*.

Prostratin proved to be a white crystalline solid that when ground up looked much like ordinary table salt. Ironically, the chemical itself had been discovered back in 1976, at which time it had been identified as "a toxic tetracyclic Di Terpene ester." Back then, the chemical's potential disease-fighting properties were unknown: this was well before the AIDS era, and prostratin had never been tested against that disease. In 1995, however, Cox, Cragg, and the other NCI lab workers filed a joint U.S. patent application describing "an antiviral composition, in particular an anti-HIV composition, and methods of treating viral infections."

In February 1997, after a two-year wait, Cox, Cragg, and the others were issued U.S. patent no. 5,599,839, for an "antiviral composition." Four years after that, in the spring of 2001, prostratin was finally licensed—not to a drug company, however, but to the AIDS Research Alliance of America (ARA), a Hollywood, California–based AIDS research group that had plans to begin testing the drug in phase 1 clinical trials with humans.

Almost two decades had passed since Cox had first heard about the *Homalanthus* preparation in the rain forests of an isolated Pacific outpost, but by 2002 the drug had still not reached the first human AIDS victim.

That, unfortunately, was par for the course in contemporary drug discovery. Clearly, there had to be a better way of finding new drugs than this.

THE PRINCIPAL DRAWBACK to the rain-forest approach to drug discovery was the length of time it took to go from the unknown jungle substance to the lab, from there to the patent office, then to a drug company, to the FDA, to clinical trials with animals and humans, and then finally and at long last, to the marketplace. The entire process could easily take twenty years or more, which was not too surprising given how unlikely it was that some odd molecule from the deep woods would have any bearing at all on human life, much less constitute a wonder drug for a human illness. The fact that the rain-forest approach ever worked at all was just barely short of a miracle. Fortunately, there were at least two other main routes to discovering new drugs.

The first was known as rational, or structure-based, drug design. Appearing in the 1970s, at about the time that Paul Alan Cox was making his first missionary foray into Samoa, the strategy worked by locating a given molecular "target" in the body, the target being the microscopic cellular structure that played a causal role in the disease in question. With the target identified, crystallographers and other researchers would try to determine its three-dimensional structure, after which medicinal chemists would attempt to design a second molecule that would fit the target in such a way as to inhibit its action, thereby preventing, alleviating, or curing the illness. The final product, assuming that all of this worked out as planned, would be a medicinal chemical, a drug that was so perfectly tailored to the original target that it had exceptionally high specificity of effect, meaning that while it worked well against the disease, it caused few if any side effects.

The rational, structure-based method of drug discovery had its share of successes. In 1975, researchers at the Squibb Institute for Medical Research used crystallographic images to design the drug captopril, which was useful against hypertension. Soon, other scientists at Abbott Laboratories were designing HIV protease inhibitors for use against AIDS, others at Biogen were at work on thrombin inhibitors for blood clotting, and still others at Sterling-Winthrop were designing a rhinovirus coat protein targeted at the common cold. To remain competitive and miss no opportunity to discover a breakthrough drug by any method, most large pharmaceutical companies would eventually retain teams to pursue rational drug design.

But for all the promise of the structure-based drug approach, the process of designing drugs to order was limited by a number of factors, the first being the fact that the molecular basis of a given disease was not always known to researchers. In such cases, drug discoverers had no idea about which disease targets to work with, so that the whole design process could not get off the ground. Second, even if a molecular target was known, its precise molecular structure might be obscure or practically

impossible to decipher, in which case it would be hopeless to try to design
a drug that would interact with it at all, much less one that would do so in
classic lock-and-key fashion. Many drug targets were proteins, after all,
and determining the configuration of a given protein was a notoriously difficult problem in applied science.

There was a second route to new drug candidates, however, one based
on a diametrically opposite philosophy. Instead of tailoring a specific drug
to a specific molecular target, the idea was to create hundreds of thousands, millions, or perhaps even billions of different potential drug molecules, essentially at random. Out of such an enormous embarrassment of
riches, the odds were that some small subset of those compounds would
be effective against one or more target diseases. For obvious reasons, this
was known as the "brute-force" (also known as "high-throughput")
approach to drug discovery: whereas rational drug design was like making
a chess move only after lots of intelligent advance planning, the brute-
force method was like trying out millions of possible moves at every turn
until essentially by accident, the proper combination of moves worked.

In favor of the brute-force approach was the undeniable fact that many
more compounds were chemically possible than had ever been tried out
against diseases. Indeed, given the total number of the chemical elements
and the laws of chemical combination, many more compounds were theoretically possible than had probably ever existed. Chemists estimated, for
example, that 10^{200} organic compounds of molecular weight less than 850
(a desirable weight for drug molecules) could be constructed: that was an
astronomical number of compounds, and almost certainly some of them
could cure human ailments.

Far-out as it seemed, the brute-force approach had been invented not
by chemists but by nature itself, in the form of the immune system. The
immune system worked by generating tiny particles called *antibodies* that
fit into the cracks and crevices of invading microorganisms in such a way
that prevented them from damaging the body. The human immune system was capable of producing about a trillion different antibodies merely
by shuffling and reshuffling their component parts. The system then
selected from among the total by a complicated process of mass screening, thereby finding the specific antibodies that were most effective
against a given pathogen.

Starting in the 1980s, at about the same time that Stu Kauffman had conceived a method of creating new proteins by the million, chemists were coming up with ways of reproducing the body's own talent for fighting diseases by creating large numbers of protective particles and directing them at the molecular enemy. One method of creating those molecules, invented in the mid-1980s by H. Mario Geysen, a chemist at Glaxo Wellcome, was known as *parallel synthesis*. Essentially, parallel synthesis worked by starting out with many small lots of a given substance and systematically adding a different new chemical to each lot. Repeating this process again and again by adding new chemicals to the mix resulted in an ever greater number of different chemical combinations.

The standard way of doing this in the lab involved using microtiter plates, small plastic trays dimpled with ninety-six separate wells into which tiny amounts of chemicals could be added, either manually or robotically. Each tiny well served as its own distinct chemical reaction vessel, and at the start of the process each of the wells received a small dose of the same starter mixture. From there it was simply a matter of systematically adding different chemicals to the wells, one after the next. For example, you could add a small amount of a known chemical to all twelve wells in the first row, a second chemical to the second-row wells, and so forth. Then you could do the same with the columns, adding one compound to the first column, another to the next, and so on. You would end with ninety-six new substances. By multiplying the number of wells or the number of chemicals deposited in them, or both, virtually any arbitrary number of new compounds could be produced at will. It was a mindless, tedious, mechanical process, made to order for robotics, and there was virtually no upper limit to the number of different chemical compounds that could be produced over time.

The first automated device for parallel chemical synthesis was constructed by scientists at Parke-Davis. The robotic unit generated a mere forty new compounds at a clip. Later, scientists at Ontogen developed a system that churned out one thousand new chemical substances a day. The products were stored in chemical "libraries," which soon contained tens of thousands and then hundreds of thousands of unknown new compounds.

In the late 1980s, a second brute-force technique was invented by Árpád Furka at Advanced ChemTech in Louisville, Kentucky. This method, known as split-and-mix, was essentially the reverse of the first. It worked by starting out with a bunch of different chemical compounds in test tubes. The experimenter then mixed them together and added a new compound, separated the result into different lots, then recombined them and mixed them with a second new compound, progressively creating an inventory of compounds whose numbers increased exponentially in each cycle. It was like taking the rollers of a slot machine, cutting them up into their individual picture-frame units, pooling all the different pictures together while adding in a new image, and then creating new rollers that incorporated the new picture. Since the new picture would be added in at different places on each of the new rollers, each roller would hold a different sequence of images than its neighbor.

The split-and-mix process could be repeated any arbitrary number of times, and as was true of parallel synthesis, there was virtually no upper limit to the number of new substances it was possible to create. Within a few years of the technique's invention, scientists at Chiron Corporation of Emeryville, California, developed an automated split-and-mix chemical combination system that turned out literally millions of new compounds in a matter of weeks.

Either way, whether by parallel synthesis or by split-and-mix, the techniques of automated combinatorial chemistry made it a simple matter to create vast stockpiles of new molecules in incredibly short spans of time. Still, the very size and variety of the stockpiles that gave the brute-force method its potential for drug discovery made for an equally great obstacle in choosing among them. With such a dizzying array of new molecules available, the problem became one of selection: How was it possible to winnow down the immense starting field to a manageable number of compounds, separating the useful from the useless, the toxic molecules from the miracle cures?

It was a data-dump problem of historic dimensions. But in 1997, an Info Mesa startup company would come up with a completely novel solution.

AFTER STARTING SOME other companies in the high-tech medical field, including one that made and marketed an automated blood-gas sensor for use with dialysis patients, Anthony Rippo, the future CEO of Bioreason, thought it was time for a rest. Rippo's father, the tuna fisherman, died in 1992, and a couple of years after that Rippo decided to quit his latest company, Unifet, leave San Diego, and as he described it to himself, "have an adventure."

He and his second wife, Madeline, had raised their six kids, all of whom by this time had flown the coop. Both mother and father had worked hard, and they needed a break from all this constant professional work and responsibility, plus the standard hassles of big-time parenting. So they decided to put their house on the market, see if it sold, and if it did, then find a new place to live, preferably somewhere other than southern California.

To their amazement, the house sold in a week. To their further shock, the buyers wanted to take physical possession of the place within two more weeks. This was duly accomplished, at which point Anthony and Madeline, although well-off financially, were homeless. So they packed their bags, got in the car, and started rambling around northern California. Tiring of this after a couple of weeks, they headed east, to Colorado. Some friends of theirs had talked up the town of Durango, saying it was exactly the place for them. But it wasn't.

They drove south into New Mexico and stopped at Taos, supposedly the artistic paradise and garden spot of the Golden West. It was pleasant enough, they had to agree, and for a while the Rippos even tried to rent a place there, but none of the landlords would accept dogs. Anthony's long-time hunting companion, a well-trained black Labrador retriever named Bella, was with them on this trip, and they were not about to part with her just for the sake of renting a house. *Adios*!

Finally they rolled into Santa Fe, where dogs were an inherent feature of the landscape and were often found lounging around in retail shops, just like in Paris. The Charlene Cody Gallery on West Palace Avenue, for example, kept a pair of West Highland white terriers in the showroom, Andrew and Piper, two little live wires who cheerfully greeted each potential customer and just possibly made a few sales.

Anthony and Madeline rented a house on East Alameda, not far from downtown. They decided they would remain there for the entire fall— they would not budge, they would not travel, they would not do anything at all but rest, relax, and enjoy the good life. Two weeks later, walking along Canyon Road, they happened upon a 130-year-old house that had a "for sale" sign at the front. In fact, it was more than just a house—it was, according to the realtor's listing, a "historic adobe residential compound," an impressive property even for Santa Fe, already full of overpriced and extraordinary pieces of real estate. The complex went back to 1862 and in more recent years had been owned in turn by an heiress, a Russian prince, and Ruth Underhill, the American anthropologist.

The prime feature of the main house was a forty-four-foot-long "grand room," a huge space that had once been a dance hall. Adjacent to it was a cantina and beyond it, a garden. In addition there was a sunroom, kitchen, and two bedrooms, all of which had hand-plastered walls, high viga ceilings, brick or wood flooring, and the original 1870s doors and windows. The second dwelling was an eleven-hundred-square-foot guest house, complete with its own kitchen, living room, bedroom, and bath, plus a courtyard dotted with age-old fruit trees. The third house was a tiny freestanding structure that had originally been an adobe stable but was now ideal for a home office, studio, or shop. This was perfect for Anthony, whose hobby was woodworking and who made concert-quality violins and violas in his spare time. Two small storage buildings rounded out the property.

Well. All of it was quite attractive. The place needed some work, true enough, but Madeline and Anthony decided that here was an opportunity that could hardly be passed up by anyone in their right mind who was flush with cash, which they were, and so they bought the complex and moved in.

They spent all of 1995 remodeling the place—power saws, tool belts, sander machines, the works. "It just about killed me," Tony Rippo said later.

Meanwhile, as was only proper for a freshly minted Santa Fean, he was having his first brush with the great spooky realities that, according to many locals, lay just below the realm of appearances. As part of their adventure, the Rippos had stopped in Santa Cruz, California, where they'd visited an alternative medicine bookstore. Anthony happened to leaf through a New Age magazine that had a story in it about one Larry Dossey, M.D. Dossey had written a book called *Healing Words*, about the intimate connection between prayer and medicine. Rippo was a practicing Catholic and not at all into the paranormal, but somewhat to his own surprise, he was intrigued enough by the magazine piece to buy a copy of Dossey's *Healing Words*.

"This was sort of like buying a pornographic book for me, buying a New Age book!" he remembered—it was on that big a scale of embarrassment. But he took it home and began to read it, "and this mind-body medicine stuff was pretty fascinating." While he was still in private practice in the 1970s, Rippo, as did other physicians, had had his share of patients who got better on their own for no discernible reason. He himself had about thirty such cases, some of whom were cancer patients. "They're sort of troubling," he said. "If you see a patient that has cancer and it's metastatic and then it all goes away, you know that there's something that caused it, but you also know that sure as hell it wasn't me as a physician. There's no really good answer—it's just something that people see."

And of course Larry Dossey just happened to be living in Santa Fe, not far from the Rippos. Anthony met him in 1995 while he was renovating the house, and the two of them discussed these weird miracle-cure cases, plus the effect of thoughts and beliefs on the immune system, as well as the possible role of prayer in recovery. For his part, Rippo told Dossey that when his own father was dying, Rippo had been appalled at the lack of heartfelt caring his father had received in the hospital, and in short order the two of them decided to host a national conference on spirituality and medicine. A year later, in 1996, the University of New Mexico Medical School sponsored the affair, which was a great and raging success, drawing some eleven hundred attendees, most of them doctors and nurses. Later Rippo and Dossey started a foundation at St. Vincent Hospital in Santa Fe teaching cancer patients that their attitudes and beliefs had a powerful influence on the course of their illness and wellness.

"At the same time, I'm just irrepressible in starting companies," Rippo admitted. He was, indeed, a regular Mr. Startup.

Rippo's next-door neighbor, John Elling, was an analytical chemist who worked at the Los Alamos National Laboratory. He'd been at the lab for seven years and was working in the automation division, teaching robots to pick up radioactive chemicals and bomb parts by remote control. Rippo, having by this time finished with both the house renovations and the spirituality and medicine conference, needed something else to do besides making violins, and so Elling got Rippo a position as a business development consultant at Los Alamos.

The lab, which was downsizing following the end of the Cold War, was pushing both its wares and its people out the door and into private industry. Rippo's job was to decide which of the lab's technologies might be most relevant to business, and then figure out how to get them out of the lab and into the world of modern commerce. He did this for a short while and enjoyed it well enough. Still, he had not started any companies recently, he was not running his own business, and so he was not entirely happy with his lot in life.

John Elling also had a consulting job with Amgen, a biotechnology firm in Boulder, Colorado. The company was heavily involved in drug discovery and faced the daunting prospect of having to wade through vast chemical compound libraries in order to identify promising new drug candidates. Elling told Rippo of the huge commercial possibilities involved in this: there were literally millions of new and unknown chemical compounds out there, tons of data sets to sift through, drug companies willing to pay millions of dollars to find the next great miracle drug, opportunities galore, and so on and so forth.

And, said Rippo, "It sounded like a business to me!"

THE KEY ELEMENTS in starting a new company, Anthony Rippo had learned through considerable experience, were few in number. The first was the general business project, the problem the business was supposed to make lots of money by solving. Then, in order, there was the proposed method for solving it, the people who were to implement the solution, and the money to pay for it all. In the case of this new drug-discovery company, the challenge was to wade through massive libraries of chemical compounds in order to find the few that might hold some promise as drugs. In the pharmaceutical industry, promising substances were known as "lead compounds," those that might guide, convey, or lead researchers toward potent new medicines. Given that there might be only a handful of lead compounds hidden away in a chemical library numbering in the hundreds of thousands or millions of different substances, this was a classic needle-in-a-haystack problem.

Rippo got a sense of the scale involved during his first visit to a chemical company, Searle/Monsanto, in St. Louis, Missouri: "They had these huge buildings filled with refrigerated vaults, containing all these things the company had been going out and collecting for the last hundred years." In addition to those "natural" chemical specimens, however, there were also synthetic compounds, thousands upon thousands of them in ranks of ninety-six-well microtiter plates, each plate holding trace amounts of almost a hundred different compounds that had been produced by combinatorial means. All told, Searle/Monsanto's chemical library, parceled out on its collection of microtiter plates, held more than a million experimentally generated compounds: a million separate physical samples held in a million tiny chemical wells. The company also had a data strip on each compound that chemically identified it, and stored that data electronically on computers. The task was to fish out from among all the others in the overall data set the small number of promising lead compounds that would yield a moneymaker. Rippo's new startup would do this, he hoped, with the help of a technology that in recent years had come to be known as "data mining."

Other than for the name, there was nothing especially new about data mining. In principle, it was simply the practice of systematically examining large lists of facts and figures in order to unearth the desired nuggets of information: looking through a phone book for all the names followed by "M.D." was a rudimentary example of the process. Data mining came into its own as a technology, however, when computers made it possible to search large volumes of data quickly and automatically for a bunch of chosen characteristics. For example, a computer could search a database for people who were medical doctors, lived in a given zip code, and had more than one telephone. It was merely a matter of writing software that was sensitive to the relevant search objects. In the 1980s, when computers began to offer large amounts of processing power, memory, and data storage at reasonable prices, credit card companies, direct-mail firms, insurance companies, and banks, among other businesses, made increasing use of data-mining techniques and honed them to a high degree of efficiency in the process.

Those techniques were easy to describe in a general way—the computer was programmed to search for patterns and for exceptions to or deviations from the norm—but the actual implementation of those techniques in data-mining systems was something else again. The software combined notions from classical statistics such as regression analysis (predicting the value of one variable by extrapolating from a set of known variables), with some highly arcane strategies and procedures from artificial intelligence such as clustering (arranging elements according to similarity of features so that underlying patterns became more obvious), and the construction of "data trees" (branching hierarchies of related data points), and so on. All of these operations were aimed at detecting previously hidden trends, patterns, and deviations from the norm in what initially had been a highly variegated collection of elements. The result was an automated ability to convert confused and chaotic "data tombs" into sets of highly organized, structured, and practical information. In went data; out came knowledge.

The techniques of data mining were soon taken up by the military intelligence community, who utilized them to separate signal from noise in endless reams of surveillance tapes, by medical doctors who wanted to detect abnormal patterns in radiology images, and by scientists, particularly astronomers, who wanted to extract substantive results from their large surveys of celestial objects. Data-mining software, for example, would make it possible to run through an astronomical data archive and systematically separate the stars from galaxies or discriminate one class of stellar object from another. In due time there would be international conferences on data mining, and starting in 1997, the field would acquire its own scientific journal, *Data Mining and Knowledge Discovery*, published by Kluwer Academic Publishers of Norwell, Massachusetts.

However, the programmers who, together with Anthony Rippo, would cofound the new company would supplement these standard data-mining techniques with an automated reasoning system customized for drug discovery. Their idea was to take the very mental processes that medicinal chemists themselves actually used when they evaluated chemical libraries for druglike compounds, and then reduce those mental processes to a set of computer algorithms that would do the same thing, mechanically and automatically. The resulting software would be able to search through chemical databases and locate the best compounds all by itself, with minimal human guidance or input. The entire task would be done by machine.

A prototype for an automated reasoning system for drug discovery had already been created by Rippo's neighbor, John Elling, and by Susan Bassett, a professor of computer science at Florida State University who had done long-distance consulting work for Los Alamos as a faculty partner for research. In fact, the two of them had already formed an informal consulting business to do some initial chemical screenings for Amgen, and their system seemed to have worked well enough. To optimize it, Bassett, who was not herself a chemist (her doctorate was in biostatistics), had gone out into the field and talked to medicinal chemists at Parke-Davis, Pharmacia, and Biacore, asking them questions about how they reasoned when they thought about making drug molecules.

That reasoning process, she learned, was hard enough for them to describe in words, much less for her to formalize in software. Chemists told her about putting compounds into families organized by various properties, about finding similarities with other drug molecules, about fiddling with structure-activity relationships, about manipulating pharmacophore element geometries, about modulating a structure to increase its reactivity and pharmacological behavior parameters, and so on.

The druglike-compound search process was somewhat like panning for gold, she learned, with the difference that drug discoverers were not looking for individual scattered nuggets but rather for whole families of similar or chemically related compounds. The chances that an isolated molecule with bioactive properties would turn out to be an effective drug were slim to nil: a large share of bioactive molecules were toxic, had unacceptable side effects, or came close to curing a disease without actually doing so. From a medicinal chemist's viewpoint, it would be much better to have a whole slew of related molecules that shared a common chemical architecture, because then you could search through the list of similar ones in hopes of finding the ideal molecule somewhere among them.

That was the goal of Bassett and Elling's automated software for drug discovery: it would search through chemical libraries and find families of bioactive molecules that shared a common chemical scaffolding, giving the medicinal chemist a continuum of similar structures to experiment with. But translating a medicinal chemist's cryptic descriptions of how he or she found classes of compounds into workable, practical software was not easy.

"That is just an ugly class of compounds, what can I say?" chemists repeatedly told Bassett about various sets of substances. Or, "I don't know how I do it, but I know it when I see it." Or, "Well, I would never put *those* compounds together as a chemist!"

Still, through trial and error and follow-up questioning, Bassett and Elling finally created a basic knowledge-extraction system for chemistry that was able to run through large and complex data sets overnight.

Anthony Rippo entered the picture just as Bassett and Elling were about to incorporate as Biomolecular Reasoning Systems and were trying to decide whether to base the company in Florida or New Mexico. Bassett had flown out from Florida to meet Rippo, and all three of them—she, Rippo, and Elling—hit it off in grand style. Rippo would come in as CEO and be responsible for raising the initial seed money, putting the corporate structure in place, and getting the company off the ground. Bassett and Elling would come in as executive vice presidents and would mastermind the actual software development.

So in the fall of 1997, Susan Bassett quit her tenured position at Florida State in Tallahassee, put all the worldly possessions she cared about into her 1993 GMC Yukon and a U-Haul trailer, and left for New Mexico. Shortly thereafter, the three of them jointly founded Bioreason in Santa Fe with $350,000 that Rippo had managed to scrape up from family and friends.

Their base of operations was located at first in Susan's rented house on Cerro Gordo, a place with a fabulous view of the mountains, the watershed, and the local Audubon Center. There, in her living room, she set up folding tables, chairs, and a bunch of computers—the classic software development site.

Bassett's job was to get beyond their initial prototype system for drug discovery. Her final objective, in fact, was to re-create, inside the computer, the very mind of a drug discoverer.

8

THE PURPLE BARGE

IN JULY 1986, after hearing about the place from Doyne Farmer, Stu Kauffman came to the Santa Fe Institute to attend its founding workshop on "complex adaptive systems," which was Murray Gell-Mann's new and improved term for what he had earlier been calling plectics. The conference was a milestone in the formation of the new institute, as it brought together for the first time several like-minded thinkers, some of whom until then had been imagining themselves as lone toilers in the complexity sea.

"The sense of ferment and intellectual excitement and chaos and seriousness and joy is here all the time," Kauffman said of the Santa Fe Institute. "Along with the sense of *I'm not alone*."

The institute was looking for staff members, and Stu Kauffman was a logical choice. Celebrity-level scientist that he had almost become by now, Kauffman had no trouble working out an arrangement with the University of Pennsylvania whereby he could spend half his time at each place and be considered a full professor at both. "It was extraordinarily kind," he said. "Not many places would allow you to do that."

So in the fall of 1986, the Kauffman family—Stu, Elizabeth, and the two children, Ethan and Merit—moved into a big house on a dirt road on the outskirts of Santa Fe. It should have been among the happiest times of their lives, and very briefly, it was.

Stu, until that moment, had led a charmed life. He had received a fabulous education in science, technology, and medicine; made a few discoveries; formulated new theories; and invented some methods that might prove important to the future of biotechnology and drug discovery. On the side, he had pursued a number of death-defying hobbies including skiing, mountain climbing, and mushroom collecting, but apart from a few minor scratches and scrapes, he had managed to avoid injury. The closest he had ever come to mortal danger had been in his teens when as an Eagle Scout, he and seven or eight other scouts had gone on a hiking trip in the California mountains, in winter.

The group had stopped at the ranger station beforehand, told the park ranger where they were headed, and then drove off toward the trailhead, which was several miles away up a narrow dirt lane. Almost there, the road was blocked by a fallen tree. Good scouts that they were, the boys studied their maps for another hiking trail that started from a different point. They switched to that one but decided they were too far from the ranger station to inform the park ranger of the change.

They hiked to a creek and camped overnight. The next morning it was snowing hard, and so they trekked to the cars and started back. They came to a fork in the road that was not on the map, and took the downhill fork. That had seemed correct at the time because after all they'd been driving along a high ridge. By the time they realized it was the wrong fork, the snow was too deep for the cars to make it back up the hill, meaning that they were stuck at the bottom.

The group was now a dozen or so miles deep in the forest, out of food, and on foot. The air was cold. The hours of daylight were few. There was no wind and no major snowdrifts, but the snow was still falling and piling up.

Stu and his best friend, Derry, decided to hike out and get help. By dusk they'd gotten as far as an abandoned farmhouse, where they camped for the night. The next day the snow was too deep for them to make any further progress, so they returned to the campsite and brought the rest of the kids up to the farmhouse. Stu found some haystacks and made some hay tea, which was what they subsisted on until the park rangers, having followed their tracks in the snow, magically showed up and rescued the group.

"That was the first time that I ever confronted death," Stu Kauffman said later.

In October 1986, just two weeks after the Kauffmans moved into their new house in Santa Fe, Merit Kauffman, Stu's daughter, age thirteen, was struck by a car and killed in a hit-and-run accident.

THE YEARS FOLLOWING his daughter's death were both a time of misery and a golden age for Stu Kauffman. In July 1987, for his "contributions to evolutionary theory, including the origin-of-life problem," Kauffman was awarded a MacArthur Fellowship, the "genius" grant, the cash value of which at that time was $290,000, spread out across five years. The juxtaposition of the two events, Merit's death and the MacArthur prize, was unsettling to Kauffman. "One of the worst things that can happen and one of the best things that can happen," he said. The MacArthur awards were "no-strings-attached" grants, meaning that the recipient had no formal responsibilities during its tenure. Kauffman simply went on with his work.

The Santa Fe Institute, where Kauffman spent increasing amounts of his time, by then had moved to the Canyon Road convent, whose spartan facilities he shared with the other faculty members: Philip Anderson, a physicist; Kenneth Arrow and Brian Arthur, economists; John Holland, a computer scientist; and Murray Gell-Mann himself, the one and only. Because he proved to be not so adept at institutional administration as he had been at particle physics, Gell-Mann was soon relieved of his title as chairman of the board. He had habitually left letters unanswered and phone calls unreturned, and in general had acted as if such trifling matters could be left to sort themselves out on their own, without benefit of his divine intervention. Murray and physicist David Pines had been made cochairmen of the science board, the group that was to chart the institute's path through the cosmos of advanced research.

The Santa Fe Institute and complexity theory were on the upswing during this period. True to its original charter and motivation, the place became a breeding ground for all kinds of nouveau theoretical constructs and approaches, a garden of Eden of complexity science, a place where many different exotic varieties of the theory sprang up and flourished within its stuccoed walls.

For complexity science, it was turning out, was no one animal; rather, each separate theorist had his or her own different concept of what the basic problem was that complexity theory was designed to solve, how that problem should be formulated and approached, as well as what specific combination of conceptual models, mathematical tools, and computer simulations were best suited to attacking it.

John Holland, for one, saw the issue as the desirability of reproducing, in artificial systems, the workings of evolution by natural selection. Nature, he thought, had been so brilliantly successful at creating organisms that prospered and grew, it had given rise to such an unparalleled variety of functional biological organs, structures, and systems, it created ecosystems that so well nurtured all the many species that lived with it— nature, in short, had been so adept at solving problems of all kinds—that humans could hardly do better than to imitate nature's own processes in their own approach to problem solving.

You solved a problem by employing an algorithm, a formalized series of steps or procedures that led in the end to the answer. Although the term *algorithm* had once been restricted primarily to mathematics, its use was later broadened to cover problem-solving strategies of other types as well. Holland merely took nature's own favored problem-solving algorithm, evolution, and applied it reflexively to the very domain of the algorithm itself. The result was the *genetic algorithm*, one whose formulas were based on genetic or evolutionary principles. Specifically, a genetic algorithm worked by progressive selection through actual trial, a winnowing out of the best procedure from among the different original candidates. The general idea was to test a given problem-solving strategy on its subject matter, refine the strategy in light of experience, try the revised strategy, improve it, try it anew, and so on. Like a precocious child, a genetic algorithm got smarter with each passing day so that sooner or later the ideally useful problem-solving method bubbled to the top.

At the Santa Fe Institute, John Holland implemented this reiterative process on computers and found that at least sometimes, the tactic succeeded passably well. (His implementation involved creating strings of symbols that "mutated" according to a rule and were then evaluated according to the degree by which the mutated string advanced the task at hand. Each new string in turn was graded for its potential ability to solve the original problem, whatever it was, and this process was performed again and again until one or more solutions emerged.)

The genetic algorithm soon became one of the fast new abstractions at the Santa Fe Institute, synonymous with a promising new approach to complexity theory. There were others. Chris Langton, who like John Holland was a computer scientist, saw complexity in terms of natural life-forms and tried to create *artificial* forms of life, systems that behaved like natural organisms but in fact were made out of materials other than flesh, blood, and bone. Mostly, these strange new "ALife" creatures consisted of electrical impulses speeding through the junctions of computer chips. As such, they did not approximate the essentials of actual living entities in even the remotest degree—not that this ever daunted dyed-in-the-wool ALife enthusiasts, who often cited computer viruses as obvious examples of artificial life. Computer viruses, however, despite their ability to spread from machine to machine much as biological viruses spread from host to host, were merely *procedures*—programs, sequences of steps, lists of instructions—rather than autonomous physical entities, and they resembled actual living things about as much as a computer simulation of a hurricane resembled a hurricane. Artificial life, in other words, was a discipline that lacked a tangible example of the species it was supposedly studying.

In September 1987, nevertheless, Chris Langton organized an artificial life workshop at the Los Alamos lab, where Langton was a postdoctoral fellow at the Center for Nonlinear Studies. (Ironically, the lab had appointed Langton a "postdoc" in 1986, five years before he actually earned his Ph.D. from the University of Michigan with a thesis called "Computation at the Edge of Chaos: Phase-Transitions and Emergent Computation.") For a field without a subject matter, the workshop was a great success.

Later, aided and abetted by a "self-organizing committee," Langton organized a second artificial life workshop at the Santa Fe Institute in February 1990, by which time he had become a so-called visiting external professor at the institute. By the time of the second workshop, artificial life had blossomed into its own separate field of study, with an established language and a proprietary set of heroes, milestones, and body of "lifelike" computer programs, but no genuine artificial living entity ever managed to crawl, walk, swim, or fly away from any of its various proceedings.

In 1997, Langton would start his own Info Mesa company, Swarm, to market "a general purpose simulation package for the investigation of concurrent, distributed systems: systems in which hundreds or thousands of autonomous agents interact with one another and with a dynamically changing environment." The company's offices, a residential house at 624 Agua Fria, were littered with larger-than-life-size models of bees.

Unlike the artificial-life hobbyists, the Santa Fe Institute's economists at least had an external subject matter to work with—the world's financial systems, markets, and populations of rational economic agents, all of whose activities the economists modeled, simulated, and made forecasts about, and then proceeded to compare with empirical reality. Noneconomists got into the act too, and during the institute's 1987 conference "The Economy as an Evolving Complex System," Doyne Farmer and Norman Packard, the childhood friends from Silver City, New Mexico, gave presentations sketching out how dynamical systems theory could be used to create computer programs that would emulate and perhaps predict the finer workings of an economy, and maybe even the stock market. (All of this would provide foundational material for their later Info Mesa firm, the Prediction Company.)

Off in his own intellectual universe, as was his wont, Stu Kauffman kept abreast of all these emerging theories, disciplines, systems, and events and even made contributions to some of them—he gave a paper titled "Co-Evolution to the Edge of Chaos" at the second ALife conference—but mainly he wended his own way down a path that led through increasingly dense thickets of theory. He and Murray Gell-Mann, the institute's superstars, did not always get along. Murray, who took a dim view of Kauffman's penchant for seeing every last phenomenon in overly "holistic" terms, started referring to Kauffman as "the doctor." Kauffman, in response, put a nameplate on his office door that said, "Stuart Kauff-Mann."

In his years at the Santa Fe Institute, Kauffman set out on some new theoretical directions of his own, one of which took narrowly biological concepts and applied them in other domains—applying the theory of evolution to the business world, for example. That was hardly a new idea: social Darwinism, which viewed societies as evolving through mechanisms of competition, selection, survival, and extinction, had a long history that went back to the nineteenth-century philosopher Herbert Spencer. For that matter, seeing commerce through the lens of Darwinian evolution was a relatively common everyday notion: the marketplace was a dog-eat-dog world ruled by competition for scarce resources, resulting in a world of survivors and failures. If organisms mutated, competed, survived, and died, so did businesses. Stu Kauffman, however, went far beyond this with the concept of a *fitness landscape*.

The idea of a fitness landscape had been introduced by the evolutionary biologist Sewall Wright in the 1930s to capture the notion of the comparative evolutionary successes attained by the world's biological populations. Fitness, in Wright's view, was not a measure of the health of an individual but pertained instead to a species, where it functioned as a measure of the members' ability to reproduce themselves. For any given species, the fitter it was, the greater its success at producing offspring— that was the kernel notion.

Since a numerical value could be attached to the fitness of a given species, it was possible to show the comparative fitnesses of many different species graphically. One could do this with a histogram, a simple two-dimensional chart in which the fitnesses of two or more species were plotted along a common horizontal axis: the fitter the species, the higher the vertical bar that represented it.

Such a depiction, however, failed to show the true interconnectedness among the many different species with which the animal kingdom was populated. A more accurate picture could be created by constructing a three-dimensional fitness surface, a fitness *landscape*, in which the mountain peaks were pinnacles of self-reproductive success and the valleys were the embarrassing low points.

Each distinct genotype, in Sewall Wright's view, was associated with a specific fitness value, and the comparative fitness values of all the world's genotypes stretched over a vast three-dimensional terrain of reproductive competence, a surface that represented the entire universe of biological organisms. It was a sort of Himalayan range of the fit and the unfit, the space of all extant genetic combinations plotted out against their overall reproductive achievements.

To Stu Kauffman, the fitness landscape was a concept wrapped in a blue flame. Here was an idea in the Kauffman mold if there ever was one: it was a metaphor, it was abstract in the extreme, and hardly anybody knew what it meant. But to Kauffman it was a lock-opener, a secret means of access to the fundamental nature of things, a wormhole into that eternal Kauffmanian subject matter, which still was, as it always had been, the understanding and elucidation of merely *everything*!

It was perfect. And so Kauffman took the idea and ran with it. He now distinguished "rugged" landscapes from "random" landscapes, and "Fujiyama" landscapes from "NK" landscapes, whatever they were, but there were many more specimens besides just these four. Indeed, once he got going there was really no stopping the man.

As he let his mind wander across these multiple mountain ranges of evolutionary suitability, Kauffman had what he thought were major insights as to how the forces underlying the processes of natural selection actually worked. The rugged landscapes, he decided, were rugged (irregular, consisting of low peaks and even lower valleys) precisely because of the many competing requirements that a given species might have to satisfy simultaneously in order to survive. For example, it might have to be smart but intuitive, big but fast, and light but strong, all at once. This would lead to a sort of all-purpose, default-mode, compromise organism, and the associated landscape would be irregular but nevertheless somewhat dim, dull, and unexciting.

"It is these conflicting constraints that make the landscape rugged and multipeaked," he said. "Because so many constraints are in conflict, there is a large number of rather modest compromise solutions rather than an obvious superb solution. There are, in other words, many local peaks with very low altitudes."

But the important take-away message about all this, he decided, was that the general principles involved had applications to many different things besides organisms: fitness landscapes explained certain features of technology, for example. Technological artifacts, after all, were compromises among competing constraints just as much as biological organisms were.

"Suppose we are designing a supersonic transport," he reasoned, "and have to place the fuel tanks somewhere, but also have to design strong but flexible wings to carry the load, install controls to the flight surfaces, place seating, hydraulics, and so forth. Optimal solutions to one part of the overall design problem conflict with optimal solutions to other parts of the overall design. Then we must find *compromise* solutions to the joint problem that meet the conflicting constraints of the different subproblems."

This meant that the design trade-offs encountered by a technological artifact could be represented by a fitness landscape just as though it were a biological species: "The standard operating procedures on the USS *Forrestal*, the linked production procedures in a General Motors plant, British common law, a telecommunications network—all evolve on landscapes in which minor 'mutations' can cause both large and small variations."

Fitness landscapes thereupon took their place in the Santa Fe Institute's scientific firmament alongside genetic algorithms, artificial life-forms, global-marketplace simulation programs, and other competing models of complexity. In Stu Kauffman's mind, however, fitness landscapes became all-embracing explanatory schema, items that would take him, over the course of time and by a strange route, out of the realm of biological theory and into the world of business.

IN JULY 1985, about a year before Dave Weininger and Yosi Taitz founded Daylight, Dave's brother, Art, had introduced Dave to a medical student he knew by the name of Dawn Abriel. Dawn was a small and trim woman with a calm manner and an engaging smile. At that time, she was living with her daughter, Gwenevere, in Claremont, "in the world's smallest two-bedroom apartment."

The day after Art introduced the two, Dave, who was an amateur astronomer in addition to being a private pilot, called Dawn and asked if she would like to accompany him to the Mojave Desert at 3 A.M. to catch a glimpse of a certain celestial object that was making a rare and brief appearance in the northern skies.

"I do not do 3 A.M. except when I am cramming for an exam," she told him. "What else do you have to offer?"

Dave offered her a flight in his Aircoupe, *Puer Æternus*, that very evening. This was such an unusual invitation that Dawn could not resist, and so an hour or two before sunset, Dave and Dawn took off from Brackett Field in Claremont, climbed out over the Pomona smog, headed west, and flew across the Los Angeles basin and then out over the ocean. The setting sun was reflected in orange glints from the waves below, Catalina Island glimmered like a jewel in the distance, and the mountain peaks on the mainland beamed softly above the purple haze—it was a remarkable evening by any measure.

They landed at Oxnard airport, where they had dinner, then climbed back into the plane, and returned to Claremont. The experience changed both their lives.

"It was love at first sight," said Dawn.

"It was love at first sight," said Dave. "We've been together virtually every day since then."

Dave was then in the process of building up his flying time and studying to get his flight instructor's certificate, a license to teach others how to fly. He knew he was up to the task the night his engine quit above greater Los Angeles.

He was taking a friend of his, Kai Hasselbach, a German scientist, for his first ride in a private plane. It was another sunset flight, near dark, and they were moving along through calm air at about twelve hundred feet, sightseeing. Dodger Stadium, a major tourist attraction, was just about to slip under the wing, when out of the blue the Aircoupe's engine coughed, sputtered, and died. Suddenly it was unnaturally quiet in the cockpit.

Any pilot worth his or her salt knows exactly what to do in such a circumstance, however, and Dave Weininger was no exception. He went through the standard emergency procedure, looking first for the nearest airports: Santa Monica to the west, too far away; El Monte to the east, same problem. His hands, meanwhile, performed the prescribed engine-restart sequence:

Check magnetos: LEFT, RIGHT, BOTH.
Master switch ON.
Mixture FULL RICH.
Fuel valve ON.
Turn key to START.

No restart.

He slowed the plane to its best glide speed, which he knew by heart, and then took a quick look at the forced-landing spots within gliding range. These proved to be two major highways: Interstate 10 and the 101 freeway, both of which were clogged with traffic.

The plane, all the while, is steadily descending.

Off to one side, Dave spots a big, empty, well-lighted street and heads for it. This turns out to be Monterey Road, a slightly uphill-downhill street in South Pasadena.

He's on low approach to this godsend when at the last minute, a car starts down the road, coming right at him. Dave flashes the plane's landing light on and off, balloons the plane over the car with his last remaining ounce of lift, and then makes a normal landing.

The plane comes to a stop. A man who's been watching all this from his front yard, and who himself happens to be a private pilot, walks over to Dave and compliments him on his smooth landing, and then offers him a beer. Dave, slightly rattled by the experience, accepts.

Police arrive, a crowd gathers, TV crews show up.

"Did you call Mayday?" one of the reporters asks him.

"I'm at 1,200 feet at night, over a populated area with a dead engine!" Dave says. "I'm going to be on the ground in about ninety seconds! With whom would I want to start a conversation, and why?"

Kai Hasselbach, the passenger, is the very image of German aplomb: "Vee vere just looking arund und Dave said vee had a slight problem . . . "

The problem was later traced to the plane's fuel system.

Dave Weininger was not exactly accident prone, but he did engage in some potentially dangerous hobbies, another of which, in addition to flying, was motorcycle riding. That activity came to a halt one night when Dawn, having received her medical degree from the College of Osteopathic Medicine of the Pacific in Pomona, was on intern duty in the emergency room of a small local community hospital. A patient who had broken his right clavicle and right knee in a motorcycle accident was wheeled in. Upon further inspection, the victim was revealed to be none other than Dave Weininger.

Much later, Dave would also have what he describes as "a minor windsurfing accident," in which he tore the sail of the craft, sustained a few blows to his stomach, and twisted his knee. The latter injury, in turn, had its own little sequela, causing him to twist his ankle while walking.

Maybe he *was* slightly accident prone. Anyway, it was several months after his famous miracle landing in South Pasadena that Dawn accepted a residency in the emergency department at Charity Hospital in New Orleans. She, Dave, and Daylight Chemical Information Systems, which was finally a going concern with actual employees, transferred its research headquarters to Louisiana. (Yosi Taitz, CEO and ruler of the business side of things, remained in California.)

Daylight moved into a converted sugar factory at 111 Rue Iberville, a place with a gorgeous view across the Mississippi. The office was spacious enough for Dave and the three other members of the research team who worked with him, but it was still rather austere—not that this mattered much to a bunch of programmers.

Dave and Dawn, meanwhile, looked around for "an interesting place to live." What they meant by "interesting," however, was not what most other people understood by the term, and at length they found themselves inspecting a series of river barges that were docked at various ports in Louisiana and Mississippi. They found one they liked, but as Dawn said later, "It required some hull work." This was no problem for her—she could supervise the repairs, expert woodworker that she was. Physically putting the shipyard workers in motion, however, was another story.

"Not only was it a dance of massive machines, having tugs and cherry pickers show up in precise sequence, but it took me weeks to figure out that my going to the shipyards as a lone female was never going to get the barge into a dry dock for repair. I grew up on the West Coast, and the good-old-boy reality of how things work in most parts of the South eluded me for quite a while."

She finally hired a man to go with her and give the orders, and the work was duly accomplished. In November 1989, therefore, after an overhaul and a paint job, a tugboat pulled the 140-foot-long purple barge to an unused section of a marina on Lake Ponchartrain.

The purple was in honor of Mardi Gras, whose colors were purple, green, and gold. "I suggested a light lavender, the color of a Bic lighter I happened to have on me at the time," Dave said. "I assumed that you could get any color, like house paint. But the only purple for marine paint was called 'safety purple.' " This was an unusually bright shade, which with the barge's seven-foot freeboard, made it a fairly prominent sight even for so gaudy a place as New Orleans. Dawn planted ten banana trees in pots around the deck—they provided the green and gold, completing the official color scheme.

The road to the marina passed by a small airfield, where Dave parked *Puer Æternus*.

"It was all very wonderful," Dawn remembered. "I felt as though I were on vacation in a magical tropical paradise every time I came home."

THE BARGE MAY have been paradise to Dawn, but her situation at work was more like a preview of hell. There was a drug war going on in New Orleans at the time, with the police and the drug sellers in a pitched battle, both sides equipped with pistols, rifles, shotguns, and machine guns, and she had to deal with far too many gunshot wounds in the emergency room—an average of eight gun-related injuries per shift.

By nature Dawn was an exceptionally controlled, contained, and peace-loving person, and she found this nightly diet of war-zone conditions more than just slightly unnerving. Often enough, merely having a gunshot wound wasn't even sufficient, by itself, for a patient to be seen, given the triage of a typical night in the emergency department. Incoming patients with a "clean meat wound" (as it was known in the trade) would have to wait their turn, sometimes for hours, while the doctors took care of even worse-off victims, those with multiple gunshot wounds to the chest, neck, or head. The very worst of it was a single night in which she treated twenty-three people with penetrating trauma injuries for a combined total of twenty-eight separate gunshot wounds, her personal record.

The unpleasantness even spilled out of the hospital. One night, Dave and Dawn, both of them major gastronomes, had dinner at K-Paul's Louisiana Kitchen, the small sixty-two-seat restaurant owned and operated by the star chef of New Orleans, Paul Prudhomme. They left the place through the backdoor and were walking toward Preservation Hall, a few blocks away through a miscellaneous gaggle of French Quarter humanity, when Dave remarked, with serene satisfaction, "Isn't this perfect? College kids, grandmothers, grandchildren—all hanging out together." Whereupon, at that instant, a drunk rolled out of a side alley, took one look at Dawn, and said, "Hi, Doc!" ("A tourist got knifed in that alley two days ago," Dawn told Dave. "Died in Room 4.")

So by the time she finished her residency program at Charity Hospital in New Orleans in 1992, Dawn Abriel had grown to hate the place and couldn't wait to leave. On the day she was handed her diploma of completion, she climbed into her Chevy Suburban, drove out of town, and never returned.

She wound up at the Indian Health Service hospital in Gallup, New Mexico—a place that had just suffered through a major hantavirus epidemic, caused by exposure to infected deer mice. Several Navajo in the area had presented themselves at the hospital in severe respiratory distress, gasping for breath, only to die shortly afterward despite all efforts to save them. Dawn worked anywhere from ten to fourteen nights in a row at the Indian Health Service, a tour of duty that often left her a zombie at the end of it. Bad as it was, it was still an improvement over New Orleans, and she would remain at Gallup for the next two years.

Dave and the other members of the research team, meanwhile, flew out to New Mexico and systematically toured the state in search of a new headquarters for Daylight Chemical Information Systems. This they found in Santa Fe, "a city of mud huts," as Dawn regarded it. Still, the town, with its red rocks and red sunsets, was visually stunning, plus it had an abundance of fine restaurants, art galleries, music, and culture—all without gunshots. Moving there would mean a 207-mile one-way commute to Gallup for Dawn, but on the basis of a single round-trip every two weeks or so, she could put up with it.

Dave and Dawn moved to an old adobe house on Garcia Street in Santa Fe. The place had heated floors, good views, and low-maintenance grounds, and was within walking distance both to the downtown Plaza and to Daylight.

Daylight's new headquarters was a two-hundred-year-old pink adobe structure at 419 East Palace Avenue, three blocks up from the Plaza. Previously the building had been a religious shelter for unwed mothers, and the place still retained a monastic feel, its heavy wooden doors carved with crucifixes and inlaid with names of the saints: San Miguel, Santa Niño, San Carlos, and so on. Inside, the rooms soon took on their own proper programmer's aura, tables laden with the usual computers, programming manuals, laser printers, and copying machine.

Here Dave and the three other programmers set themselves to the task of making computers perform feats of chemical reaction processing, which was to say, getting them to do chemistry without chemicals. Daylight's whole mission in life, after all, was to reduce chemicals to information and chemistry to an information science. The company was already making money, its software systems now selling for anywhere from $2,000 to $60,000, depending on the purchaser, the system, and the licensing terms. Still, the company's product line was far from complete.

Dave's current challenge was in essence to reduce chemical compounds to numbers and then instruct the computer to calculate what would hap-

pen when you put together any given combination of chemicals. The ultimate objective was a program that would allow a chemist to enter into a
computer the SMILES for the various starting materials, the reaction
medium such as a catalyst or solvent, plus a set of initial conditions: the
quantities of the chemicals participating in the reaction; the duration of
the reaction in seconds, minutes, or hours; the temperature; pressure; and
every other relevant parameter. Then, simply press <ENTER>, and the
computer would do the rest, figuring out what the specific reaction products would be, doing all of it without fooling around with any chemicals,
test tubes, or other physical apparatus.

If, for example, you typed in the SMILES for ethanol, CCO, and the
SMILES for acetic acid, CC(=O)O, and then pressed <ENTER>, the computer would come back immediately with the SMILES for the correct
reaction products: O and CC(=O)OCC, which were, of course, water and
ethyl acetate. If actually performed in lab glassware, the experiment
would require the chemicals themselves, preparation time, plus whatever
tests it took to figure out what the end-products were, whereas by use of
Daylight's chemical reaction software, the entire wet-lab procedure was
reduced to a single line of text,

$$CCO.CC(\_O)O \jmath \jmath O.CC(=O)OCC$$

(ethanol plus acetic acid yields water and ethyl acetate), and required all
of a split second of computer processing time.

That, at least, was the plan. The execution, however, was slightly more
daunting. Atoms were invisible entities with several associated physical
properties, such as atomic number and weight, charge, and the number of
attached hydrogen atoms. Molecules were combinations of atoms, and
their ability to combine with other molecules was governed by a variety of
factors such as bond angles, bond strengths, valence, the number of electrons available for bonding, chirality, and many other complexities, com-
plications, exclusions, and exceptions. For all their inherent abstruseness,
however, chemical reactions were firmly governed by known rules, which
meant that reducing chemical reactions to computer procedures was just
a matter of capturing those rules in a formal language, installing the code
in the central processing unit, and getting it all to work reliably on the
input data.

Rule governed though it was in principle, it took several years before Dave and company got their reaction processing program to work smoothly, quickly, and correctly. After all, they had to make the computer predict reaction products not only for known chemicals on the basis of standard reactions but also for reactions involving substances that as yet were unknown to chemistry, made up of molecules never before seen, and processes that so far were untried, undiscovered, and unrecorded in the chemical literature. Unless it could do all that, the program would be useless as a tool of discovery. In addition, to be realistic and faithful to the laws of chemistry, the program had to recognize when a proposed molecule or reaction was improbable or impossible.

One of the advantages of doing chemistry in the computer, on the other hand, was that unlike in real life, where "you cannot unscramble an egg," a computer could make a given chemical reaction proceed either way, backward or forward. The reaction could go equally well from starting points to outcome, or from the outcome back to its starting points.

What this meant was that if you knew what your desired reaction product was but didn't know how to synthesize it, the computer would figure it out for you. You'd merely enter the reaction product's SMILES into the computer, and the software would provide a list of the necessary chemical ingredients, along with the precise reactions that would govern the entire transformation.

In fact, even if you didn't know exactly what a given chemical was, Dave's reaction processing program could nevertheless tell you how to make it.

9

THE DOCTOR IS IN

ANTHONY NICHOLLS'S EXPERIENCE with the Grasp program had soured him somewhat on the much-ballyhooed "academic lifestyle." He continued to collect royalties from his optimized DelPhi II software, but as for Grasp, which was largely his own creation and was a wild success so far as actual usage by the community of protein chemists was concerned, he'd never received a cent. The resulting lack of adequate folding money was reflected in his and his wife Teresa's living conditions. They'd settled themselves in the furthest reaches of uptown Manhattan, on 168th Street, in a block of apartments owned and operated by Columbia University. The place was a cross between graduate student residences and public housing, but as far as they were concerned, it was more like living in a Third World combat area. Two years later he and Teresa moved to the main Columbia campus, a couple of miles downtown. "That was only mildly depressing," he said.

Teresa, meanwhile, was having problems of her own with the graduate assistantship that provided her with work and income. She'd started out in Charlie Cantor's lab on an electrophoresis project but had not enjoyed it much. Later she did crystallography work on DNA-protein complexes in the lab of an assistant professor who ended up never getting tenure at Columbia—another dead end. And then in 1994, Nicholls himself had a falling out with Barry Honig over the commercialization of Grasp.

"It was basically over money," said Nicholls. "You know, who was going to get what."

Anthony had already started looking for another academic position, but his relative deficiency of conventional publications put him at a disadvantage in an academic job market ruled by the publish-or-perish syndrome. With most of his time spent writing computer code, Nicholls had little energy left over for writing formal scientific papers. Privately, he felt that his software systems were *his* publications—especially Grasp, which he regarded as by far the best piece of work he'd ever done—but his software-as-publication viewpoint was obviously not shared by those who did the hiring at college and university biology departments.

"I was enabling people to do science and to look at things in a new way, but it made no difference," he said much later. "I still have a stack somewhere of all the rejection letters I got. Not only did I get rejected, they weren't even inviting me for interviews.

"It was clear that what they wanted were people who were doing lattice simulations of proteins, one of the most useless, unproductive, unfruitful endeavors in the entire history of molecular modeling," he added. "But if you did that you got a faculty position!"

So Nicholls started applying to drug companies, big ones on the order of GlaxoWellcome and the like.

It was at about this point that Nicholls went off to Albuquerque to demonstrate Grasp at an academic conference. The software generated protein surfaces so quickly and realistically that people wanted to know if he'd created the images beforehand and stored them in a file for demo purposes. No, he told them: the software was producing the molecular surfaces from numerical data in real time, just as they watched. Viewers found this amazing.

But an even more momentous event occurred in Albuquerque, for this was where Anthony Nicholls first met Dave Weininger. Dave was at the same conference, demonstrating the Daylight product line—SMILES, ClogP, Thor, Merlin, and some others. Nicholls was charmed by the speed, power, and sophistication of Dave's systems, along with their practical, everyday usefulness to working chemists.

Of equal personal interest to Nicholls, however, was the fact that Weininger had made the big jump out of academia and had survived. In fact, he seemed to be quite well-off, at least so far as Nicholls could tell. And Weininger wasn't working for some huge conglomerate either; instead, he had struck off on his own and, together with Yosi Taitz, founded an independent business that not only had created some fabulous new specimens of scientific software but also was selling them to chemists for what to Nicholls were vast sums.

Nicholls and Dave met up next at the CEX (Chemical Exchange Language conference), sponsored by Glaxo and held at Research Triangle Park, near Raleigh-Durham, North Carolina. They ran into each other again at the American Chemical Society meeting in the summer of 1995, at which time Anthony told Dave that he was having a hard time deciding whether or not to take a job with Glaxo, where he'd had a successful interview. "I asked for more than the senior person interviewing me made," Nicholls said later. "He gulped but agreed."

Still, working for some anonymous drug company wasn't really what Nicholls wanted to do with any major part of his life. He wanted to write mind-blowing software, stuff that would outdazzle even those who'd been amazed by Grasp.

Anthony and Dave also kept in touch by e-mail. The more Anthony thought about what Dave had done, the more attractive it seemed to follow in his footsteps and strike out on his own. On the other hand, the offer from Glaxo was still on the table, as was his current assistantship at Columbia.

In September 1995 he put all this in an e-mail to Weininger and asked him for advice. Two weeks later, in a reply that ran to six full pages, Dave sized up the situation as he saw it.

"Frankly, I can't see you doing the corporate thing just yet," Dave told him. "It's not that you wouldn't do well—you'd do fine, I'm sure—it's more that I don't think you'd thrive in that environment. It would be different if you needed a seriously huge amount of support, e.g., a cyclotron or a supercooled NMR or a team of crystallographers or something like that. Then you'd have to accept some compromises to get your work done. But as you know, world-class software can be written in a very modest environment. Modest in the sense that it doesn't cost too much; not so modest in other senses, like time and space and single-mindedness."

If he went to work for Glaxo, Dave said, Nicholls would be at their mercy: he'd have to work on their pet projects, whereas what he really wanted to do was write his own private dream software. "Glaxo is not a bad choice," he said, "but might be suboptimal if your goal is to create a new way of looking at molecules rather than designing profitable drugs for a big company. What's the point of doing great things and having a terrible time?"

As for Barry Honig and Columbia, "I'd say, 'Get out.' I don't know exactly what advantages being at Columbia offers (I assume that there are some), but they would have to be pretty profound to keep me in a position where I couldn't control things of primary importance to me."

And finally, in a section entitled "WHAT YOU SHOULD BE DOING," Dave told Nicholls to dump Honig, quit Columbia, forget about Glaxo, and start his own software company. "You should formalize your 'algebra of surfaces,' make a list of what fundamental algorithms need to exist, and divide them into four categories: (1) you already know how to do it, (2) you haven't done it yet, but can do it in a reasonable amount of time, (3) you can't do it, but someone else can, and (4) it can't be done without a major breakthrough.

"You should then create reusable modules which implement these algorithms, each solving a specific problem. You should then create a viewer to express your ideas visually."

Dave also knew that Anthony was fond of New Mexico, particularly Santa Fe, and that Nicholls had decided way back in 1987, during the five-week-long Matrix of Biological Knowledge Workshop, that if he ever had the option of living absolutely anywhere in the world he wanted, Santa Fe would be the place.

"You get something different for your money here," Dave wrote. "I'm watching the sun come up now in a clear sky, peeking through the piñons in the high mountains. Not everyone who lives here is a movie star or president of a company (though when I went to the movies the other night, Steven Spielberg was sitting two rows in front of me!). Car mechanics and barbers live here too. You won't starve to death, but you probably won't buy a house right away, either.

"If you're frustrated with your current circumstance, do something about it," he added. "Life's too short for any other options. If security is an issue, try Glaxo (but not before looking around). If you want to have a crack at a dream, go for it. In my opinion, the risks aren't that great.

"You should either decide to really do it, or not. I personally think you should do it."

IN 1993, AFTER seven years at the Santa Fe Institute, Stu Kauffman published his first book, *The Origins of Order: Self-Organization and Selection in Evolution*, issued by Oxford University Press.

Even for Stu Kauffman, *The Origins of Order* was deep. It was a dense, abstruse, technical account that pretty much summed up Kauffman's professional work: his theoretical research agenda, experimental results, and scientific speculations up to that point in his life (he was fifty-four). Bursting with tables, charts, matrices, and equations, along with representations of miniecosystems, graphs depicting coevolutionary avalanches, extinction events, and more, and weighing in at 709 pages, the book was a challenging reading experience even for a scientist, and all but out of reach for any but the most committed and determined layperson. Still, endorsed on the back cover by two Nobel laureates plus the science writer Stephen Jay Gould, the book sold well enough.

It was somewhat of a surprise to Kauffman, however, when he learned from a friend of his what the true meaning and message of his immense and erudite technical tome was. The friend was Larry Wood, who was then senior scientist at GTE Government Services in Needham, Massachusetts. Wood had been hired by GTE in 1979 to work on "intractable problems," problems that no one else had been successful in solving. He had made considerable progress on such matters by viewing things from radically new perspectives, a talent that had served him quite well over the years. Wood was of a highly independent turn of mind, to say the least, and between 1995 and 1998 he and his wife, Lisa, a former advertising executive in the publishing industry, had moved to Santa Fe and set up a private quantum mechanics research lab in a renovated building in the foothills of the Sangre de Cristos. This was not a garage or basement operation—it was a full-fledged, self-contained facility housing a Beowulf supercomputer, experimental physics equipment, and a staff consisting of scientific, engineering, and administrative personnel, plus a separate security crew, all of it funded by nobody but themselves. Later still, Wood and his wife would close out that project, move to New Jersey, and start their own biotechnology company, Sunyata Systems, devoted to "precision drug design."

Anyway, one morning in late 1993 or early 1994 Larry Wood was sitting in his office at GTE and doing some exploratory reading in a volume of the Santa Fe Institute's "Studies in the Sciences of Complexity" series when he came upon a paper written by Stu Kauffman on the subject of fitness landscapes, those evolutionary peaks and valleys that Stu was so inordinately fond of. Larry Wood loved the fitness-landscape metaphor, which was new to him, and immediately placed a call to Kauffman, whom he had never met and didn't know in the least, to find out more. Placing such calls was a habit of Wood's—he always went straight to the top, to the highest authority he could find on any subject, no matter what and no matter who.

Kauffman, as it happened, answered his own phone and was more than happy to talk to Larry Wood about fitness landscapes or anything else. Stu Kauffman suffered from a well-known willingness to talk to anybody about anything, ad infinitum, or at least until the listener was ready to bolt from the room screaming.

At one point in their phone conversation, Kauffman mentioned his new book, *The Origins of Order*, and afterward Wood went off to buy a copy. Soon thereafter Stu came to Boston to give a talk about fitness landscapes at Arthur D. Little, the Boston consulting firm. Larry Wood, who always kept abreast of whatever was happening in his immediate geographical area, was there in the audience. He was also at the dinner afterward, held in the executive dining room.

There they were, all sitting around the dinner table—Kauffman, Wood, and the higher-echelon theoreticians of Arthur D. Little—when during a lull in the conversation Larry Wood blurted out, "You know, Stu, *Origins* really has nothing inherently to do with biology. What it's really about is business and management."

Even Stu Kauffman, who normally could not be flapped by anything, was fairly taken aback by this. In addition, there was a sense of profound shock all around the dinner table, as if Larry Wood had committed some highly embarrassing social blunder. Anyone who'd opened a copy of *The Origins of Order* could immediately see that it was about evolutionary biology—albeit on an excessively abstract level—from cover to cover.

But Larry Wood went on to give the basis for his extraordinary new interpretation of Stu's book and its contents. The book was so very abstract, he said, that while reading it he imagined what it would be like to try to understand the subject matter if he were from another planet. In other words, if he were coming at the book from the standpoint of an intelligent being from outer space, one who knew nothing whatsoever about biology, especially earthly or human biology, then what would he think *The Origins of Order* was about?

Wood's answer was, it was about organizing things, it was about building systems that were robust and complex and that didn't break down easily, which was to say, it was about the organization of business.

"The broader context of the book, to me, is clearly about how to analyze business dynamics, and methods for improving them," said Larry Wood.

It took awhile for Stu to take this seriously, much less to accept it, so entirely new and radical was the idea. In time, however, the notion would start to burn a hole in his head: all that stuff in the book about "sinusoidal transcription and protein patterns," about "spatial harmonics suggested by mutants," about "longitudinal deletions and mirror-symmetric duplications," and whatnot, wasn't really about biology after all.

It was about business.

IN MARCH 1995, Chris Meyer, an analyst with the Boston consulting firm Ernst & Young, heard Stu Kauffman give a talk at a conference titled "Complexity and Strategy: The Intelligent Organization," sponsored by the Santa Fe Institute and held on the West Coast at the Mark Hopkins Hotel in San Francisco. The conference fee was $3,000, not unusual for a high-level corporate caucus of the gurus, which this one gave every indication of being: not only was Stu Kauffman on the program, so also were several of the other superstars of complexity theory, including Chris Langton (Mr. Artificial Life), John Holland (Mr. Genetic Algorithm), and Murray Gell-Mann (Mr. Complex Adaptive Systems).

Chris Meyer, however, was in a better position than many others in the audience to understand and appreciate the full import of the proceedings. He had been a dual major in mathematics and economics at Brandeis University and had been following cutting-edge science ever since. "If you want to know the future of business, you look at what's going on in science, because that's one major source of innovation," he said. "Just like if you want to know the future of markets, you look at birth rates, because they tell you a lot about the market segmentation of the future."

Accordingly, Meyer had been tracking all the latest science fads: catastrophe theory, chaos theory, and now finally complexity theory, especially as it had been coming out of the Santa Fe Institute, and most especially as it had been coming from Stu Kauffman, whom Meyer regarded as the holy father and supreme master of the science. Why couldn't the laws of complexity theory be applied in the business setting and utilized for the purpose of solving practical and real-world business problems? Meyer wondered. Complex systems were complex systems: it made no difference to the final outcome whether the elements involved were natural or artificial.

This was pretty much Stu Kauffman's own general message as he stood up to give his lecture and talked about organisms and organizations, the evolution of complex systems, fitness landscapes, and all the rest of it. He also drew some explicit parallels between biology and business, saying things like, "Organisms speciate and then live in the niches created by other organisms. When one goes extinct, it alters the niche it helped create and may drive its neighbors extinct. Goods and services in an economy live in the niches afforded by other goods and services. Or rather, we make a living by creating and selling goods and services that make economic sense in the niches afforded by other goods and services. An economy, like an ecosystem, is a web of coevolving agents." And so on.

Kauffman's lecture was highly suggestive to Chris Meyer, at least the parts he could grasp and assimilate. "I probably understood two-thirds of it," Meyer recalled later. "Throughout, I had a tremendous excitement about the applicability of this material to social systems. In their physics envy, social scientists had tried to apply the tools of linear analysis to social systems, but nothing is less linear than social systems! But when I saw Stu, . . . "

When he saw Stu, everything changed. In his approach to the modeling of nonlinear systems, in his descriptions of the way agents actually evolved in dynamic, rapidly changing environments, Stu seemed to have captured the very point and essence of the market system. More, he seemed to have formalized on an abstract, theoretical level the ways in which people in business situations actually behaved.

Stu's talk ended just before lunch. There were only two elevators to the Top of the Mark, the hotel's restaurant, and Chris spotted Stu waiting in line. Chris adjusted his own position in the crowd so that he would wind up next to Kauffman as the several lines converged. The two of them spent the next half hour talking about the business applicability of complexity theory. "Clearly," Meyer remembered, "Stu was quite interested in the use of his concepts in the real world of business."

They talked again a few months later, by which time Chris Meyer was director of Ernst & Young's Center for Business Innovation. His mission there was to identify promising new business concepts and capabilities, and to acquire them for the exclusive use of the parent company. Meyer was convinced that Kauffman had a lot to offer them, and speculated that perhaps Stu and Ernst & Young could form some sort of partnership. If complexity theory was really all it was cracked up to be, then why couldn't it be applied to the complex system of modern business practice, yielding some rather stellar practical results? Businessmen would bring their problems to Stu, and he, using his unique approach to complex systems, would solve them—for a price, of course.

Meyer proposed such an arrangement to Stu, who, to put it mildly, expressed considerable enthusiasm for the scheme.

Meyer, however, now had the formidable task of convincing the chairman of Ernst & Young, Roger Nelson, that his company should enter into a business agreement with the founding father of complexity theory. Meyer wrote up his arguments in a nine-page confidential memo ("Subject: Major Thought Leadership Opportunity: Complexity and Stuart Kauffman"), and on June 28, 1995, he submitted it up through the ranks.

The advantages of forging an alliance with Stu Kauffman, Meyer claimed in his memo, were four in number: it would give Ernst & Young exclusive access to a leading-edge thinker in an important business area; it would associate the Ernst & Young brand name with innovative thinking; it would create new software products for their own use and for sale to others; and it would deny their competitors access to these same key resources. That was his battle plan. In addition, Meyer suggested that Ernst & Young purchase fifteen hundred copies of Kauffman's forthcoming book, *At Home in the Universe*, and send them to their top clients, essentially as a marketing tool, as a way of drumming up business for the proposed partnership.

Basically a popularization of *The Origins of Order*, the new book also made some pointed references to the relevance of complexity theory to the conduct of business. Biology was no longer Stu's exclusive universe of discourse.

Roger Nelson was convinced by Meyer's memo, and in short order

Ernst & Young ordered a special edition of *At Home in the Universe* from the publisher, Oxford University Press. The books would be printed with a presentation strip across the dust jacket telling the recipient that it was from Ernst & Young. There would be a cover letter inside explaining to those who received a copy that this stuff was the wave of the future.

Separately, Chris Meyer also proposed that Ernst & Young sponsor their own set of conferences on the topic. In July 1996, therefore, "Embracing Complexity 1," the first of five such gatherings, convened in San Francisco. (Others would be held in Boston and Paris.)

Somewhat to Chris Meyer's surprise, the list of speakers that he himself put together revealed that if anything, he and Stu were slightly behind the complexity-theory-and-business curve, for there was apparently a considerable list of early adopters out there already. General Motors was using an agent-based artificial intelligence system whereby robotic sprayers collectively created their own truck-painting schedules, making for savings on paint and producing higher-quality work in the bargain. A firm called Excalibur Technologies was using genetic algorithms to "breed" software for fingerprint recognition. Seiichi Yasakawa, the head of Yasakawa Electric, gave a presentation titled "Complexity and the Japan Central Railway," describing how the company had applied complexity theory to train scheduling.

Even the United States Marine Corps, it turned out, was actively pursuing complexity theory! This became clear from a talk by Lt. Gen. Paul K. Van Riper entitled "Corps Competence: How the Marines Simulate the Chaos of War."

"We are out of money; therefore, we must think," he quipped.

The Marine Corps, Van Riper said, was studying the applicability of some complexity theory–based metaphors, policies, and tactics to the conduct of warfare, an enterprise that was arguably a highly nonlinear phenomenon. "Intuitively, most of us already felt that land combat is a complex adaptive system," he confided. Combat was in fact not so much like a game of chess, which was the classic military model, as it was like a collection of individual agents (soldiers) operating on their own volition under a system of general rules (the rules of engagement), the very paradigm of a complexity-science approach to the way things worked. Complexity theory, perhaps, could help military leaders develop better rules. "We want to evolve the lowest-level rules," he said, "to look at neural nets, to discover otherwise unseen behavior patterns on the battlefield."

Kauffman's own presentation, "What's Under the Hood: A Layman's Introduction to the Real Science," illustrated by a bunch of coffee-stained slides, was a concentrated, hour-long digest of his theories to date. All of the major Kauffmanian concepts put in their respective appearances and strutted their time on the stage: fitness landscapes, ordered regimes, chaotic regimes, coevolution, conflicting constraints, Boolean hypercubes, tyroidal lattices, Glauver dynamics models, and assorted other equally esoteric notions. It was easy to get maxed out on the abstraction of it all, listening to him. And what any of it had to do with the conduct of business (the ostensible subject of the conference) remained something of a mystery for the duration of the lecture—even to Stu Kauffman, as he at one point half-jokingly confessed.

Toward the end, nevertheless, he did manage to throw out a few spare hints, enigmatic though they were. "In any production system—whether it's a job shop scheduling problem, a complex systems and manufacturing problem, or so on—they all have complex, rugged landscapes," he said. "So, what we can attempt to do is to uncover the structure of those landscapes, find out their statistical properties, and find out what the real world looks like."

Well, it was better than nothing. Most of this was so very baffling you had the impression that anyone who understood it all (as Kauffman certainly seemed to do) would be able to solve mere business problems using no more than 10 percent of his brain wattage, blindfolded.

The whole package was quite attractive to Ernst & Young, at any rate, and by the end of 1996, the company put up $6 million, Kauffman put up himself, his brains, and his theories, and the two entities formed the BiosGroup, of Santa Fe, "a partially owned subsidiary of Ernst & Young, LLP." The BiosGroup, a news release said, would use data mining, "agent-based modeling" (simulation), and "optimization" (improved algorithms) to solve a variety of business problems. "At Bios Group, we apply findings from complexity science including principles of self-organization, emergent behavior, and distributed control to create robust, adaptive approaches to management."

The whole of Larry Wood's offbeat interpretation of Kauffman's life and work, proposed barely two years earlier and seeming so bizarre at the time, had now been converted into a practical reality.

The company headquarters would be at 317 Paseo de Peralta, an adobe office building three blocks north of the Plaza. Bios needed a bunch of bright young complexologists, a cadre of mathematicians, programmers, physicists, biologists, economists, and computer scientists, all of them versed in the principles of complexity science, so Stu sent the word out, made maximum use of his personal contacts, and placed want ads in *Science*, *Nature*, the *Wall Street Journal*, and on the BiosGroup's Web page. Soon he had assembled a core crew.

At the outset, Kauffman and his group occupied a small corner of the first floor, taking up only about 25 percent of the available office space. Stu himself moved into a small first-floor room with a view of the Santa Fe post office across the street, filled his shelves with a selection of extremely abstruse science books, put his diplomas on the wall as if this were a doctor's office (which in a sense it was), and next to them, hung a framed copy of a *Scientific American* piece of his that he was extremely proud of ("Antichaos and Adaptation," 1991).

Having been immersed for the last thirty years of his life in the arcana of autocatalytic networks, the problem of cell differentiation, the genetic regulatory systems of prokaryotes, and schemes for producing millions of novel peptides, polypeptides, and proteins—not to mention the even greater mysteries of rugged multipeaked fitness landscapes—Stu Kauffman now sat back in his chair, put his feet up on the desk, and waited for the CEOs of sickly businesses to beat a path to his door.

The doctor was in.

A MILE OR so away from Bios, on Cerro Gordo Road up in the hills, Susan Bassett's house became a regular Silicon Valley–style hacker's den, filled with newly hired Bioreasoners drinking endless cups of coffee and cans of Coke, subsisting around the clock on whatever happened to be in the kitchen at the time, and napping on the couch while their programs were running. They had everything they needed right there in the house, so why go home? Bassett, herself a computer programmer of long-standing, played her part in suitable fashion, often working long past midnight in her pajamas.

She got tired of the routine soon enough, however, and Bioreason moved to a commercial office on Johnson Street in the heart of the city. The place was owned by actor Gene Hackman's wife and had once been an art gallery.

Tony Rippo now drew into the operation a major new talent by the name of Ruth Nutt. Nutt was a medicinal chemist who had spent thirty-one years at Merck and Company, where she ended up as a senior scientist and had played a key role in developing Crixivan, a pioneering AIDS treatment. She had just recently retired to Santa Fe, where she was playing a lot of tennis but not doing much of anything else. Rippo enticed her out of retirement and convinced her to join the company.

Nutt was the answer to Susan Bassett's dreams. "We now had the possibility of doing expert knowledge elicitation in-house," said Bassett, "and to build the algorithms that best captured the pieces of that thinking in computationally relevant form."

This was Bassett's polite way of saying that they were going to put Ruth Nutt's brain into the computer, along with the brain power of all the other medicinal chemists they had interviewed. John Elling, in keeping with this image, at one point walked into the office carrying a realistic-looking rubber brain that was floating around in a pool of water inside a clear plastic enclosure set on a pedestal. Bobbing around in there like a jellyfish, bathed in a lurid red light from above, bubbles coming up around it from below, the brain emitted a low humming noise as if it was processing data at a fast clip. Ruth Nutt, who spoke with a German accent and was somewhat reserved in manner, was not especially fond of this artifact. The very idea that you could simulate a brain—especially hers—either physically or in software, was distasteful to Ruth, so Elling carefully put the fake brain out of sight behind the door to his office.

Bioreason's main new software product owed a lot to Ruth Nutt, however, and for press consumption, Tony Rippo jokingly made the claim that they themselves referred to the program as "E-Ruth" or "E-Nutt." In truth, their own internal name for their first major algorithm was the equally distasteful PHLM, which stood for *Phylogenetic-Like Mesh*, a method for generating treelike classes of chemical compounds. After they submitted it for a patent, all that remained was putting it to an effective and challenging real-world test.

In 1998, Rippo had made a deal with Jim Bristol of Parke-Davis, whereby in return for its use of Bioreason's software, Parke-Davis would send them an old set of completely analyzed chemical data. Parke-Davis had screened a portion of their chemical library in 1992 and had determined the chemical structures of the various compounds within it. What they hadn't known, at least at the beginning, was the reactivity of those compounds or, most important, whether any of them possessed the magical "druglike properties," the type of bioactive chemical behavior that was the Holy Grail of modern drug discovery. Traditionally, it was the province of individual human chemists working in wet labs to make this determination, and Parke-Davis's own medicinal chemists had worked on the screen for years, evaluating the compounds one by one, finding lead compounds as best they could by old-fashioned, tried-and-true manual methods. By 1998, the data screen was ancient history as far as Parke-Davis was concerned, its researchers having decided long since that they had pretty much extracted from it whatever possible use it had contained.

"The bottom line was that they felt they knew everything there was to know about that screen," said Tony Rippo. "It was a screen they had six years' experience with."

So Parke-Davis took the raw, unanalyzed data that their own chemists had spent years analyzing, and sent it to Bioreason over a secure telecommunications line. The data consisted of a long list of chemical compounds, together with two-dimensional structural information for each item, all of it expressed in the SMILES language, for by this time Dave Weininger's SMILES notation had become an industry standard for the encoding, storage, and rapid transmission of chemical data.

The Bioreasoners now received the data screen in Santa Fe, downloaded it to their computers, and put "E-Ruth" to work on the problem.

Then they sat back to await results.

PART THREE

RESULTS

10

EYE CANDY

IN THE LATE fall of 1995, not long after he received Dave Weininger's e-mail telling him, "You should either decide to really do it, or not. I personally think you should do it," Anthony Nicholls took the plunge. He said good-bye to Barry Honig at Columbia, packed his bags, left the secure and coddled academic world that had been his home for most of his life, and moved to Santa Fe. (He and his wife Teresa had divorced about a year earlier. She moved to work at the National Institutes of Health for a while and then wound up as a patent examiner.)

He settled into a three-room apartment at 335 Winische Way, a small and obscure dirt street in the shadow of the state capitol building. Despite its proximity to the seat of government, the place seemed to be on the wrong side of the tracks. The general area had a worn-out look, with the neighborhood awash in rusting cars, stray dogs, and swirls of dust. His apartment, in the traditional one-story pink adobe building, looked out on the rear of a house that itself was not a model of tidiness. This was not palatial digs by any stretch of the imagination.

Nevertheless, it was not the slums of upper Manhattan, either, and from a block away, at the corner of Paseo de Peralta and Acequia Madre, Nicholls could look across to the Sangre de Cristos and see piñons, fir trees, and snow-capped mountain peaks through air so clean, clear, and bracing that it was a pleasure just to stand there and breathe it in. His apartment was within walking distance of Daylight's research headquarters on East Palace Avenue, Dave and Dawn's house on Garcia Street, plus a selection of fine coffee houses, where as a proper Englishman, Nicholls could even get tea.

He spent much of the first year in Santa Fe thinking about rewriting Grasp, his program for visualizing protein molecules, and worrying about making money. Optimizing Grasp, he finally decided, would not represent that much of an advance for him personally; certainly it wouldn't be the breakthrough work that he had envisioned himself doing all along. As for money, he had enough to bankroll a year or so's worth of original research in science, which was why he'd come to Santa Fe in the first place.

A series of discussions with Dave Weininger led Nicholls to consider working on software for small-molecule chemistry. Small molecules, those under a molecular weight of 500, were what drug companies were mainly interested in, since they were the class of compounds from which the most successful drugs had emerged in the past. Nicholls resolved that if he could do for small-molecule organic chemistry what he'd already done for proteins with Grasp, he would be making his own modest contribution to medicine. Grasp had given protein chemists the ability to visualize the three-dimensional surfaces of protein molecules, and to imagine how correspondingly shaped molecular structures might fit snugly into their cracks and crevices, immobilizing them or otherwise altering their biochemical activity. There were already lots of programs that gave medicinal chemists similar abilities for small organic molecules, but there were none that allowed chemists to search through libraries of such structures on the basis of complex three-dimensional properties.

Even Daylight's own search engine, Merlin, employed a two-dimensional approach to what was in fact a three-dimensional phenomenon. Merlin searched through molecules on the basis of bond structures as represented in traditional two-dimensional chemical diagrams. GRINS, Daylight's graphical input program, likewise depicted chemical molecules only two-dimensionally, not three-dimensionally. What Nicholls was after was essentially a search engine that worked on the basis of a molecule's true, real-world, three-dimensional structure in all its inherent complexity. What was important for the biological reactivity of a given drug compound was not so much the chemical bonds that held it together, but rather the shape of its surface structure, the fact that the molecule's outer surface presented the right configuration at the right place to interact with another atom, molecule, or cell receptor. Nicholls wanted to give drug discoverers the ability to search through chemical libraries for molecules of any arbitrary three-dimensional shape, more or less as if they were trying to find a glove to fit a hand. Furthermore, he wanted his software to perform these feats with the speed and pithiness of Daylight's own classic systems.

In 1997, using money that he'd saved out of his DelPhi II royalties, Nicholls founded OpenEye Scientific Software. He chose the name because he regarded the eye as a window into the brain. Since the optic nerve ran directly from the retina to the brain itself, looking into the open eye of another person was like being in direct visual contact with that person's nervous system. That was how he wanted his software to work: with the flash and dazzle of brain waves.

OpenEye's headquarters was Anthony Nicholls's living room, a large space holding an assortment of computers, books, folding chairs, and potted plants. He put up a passably cute Web page (eyesopen.com) that featured a deliberately ironic corporate motto ("Software just like Mom used to make"), outlined the "ideological background" of his work, and announced his formal mission statement: "The goal of OpenEye is to provide tools to address the explosion of chemical data generated experimentally by focusing on those features which carry the most physical importance, i.e., the electrostatics and shape of molecules, and which can be calculated at speeds sufficient for the analysis of large chemical libraries."

On a practical level, his idea was that the user would be able to input the SMILES or other identifier for a given compound, and the program would calculate the molecule's electrostatics, its three-dimensional structure, as well as practically every other relevant physical property that pertained to it, and would then store the information for later searches. Nicholls soon put together an interim program called ZAP, an application that calculated the object's molecular electrostatics alone, and showed it around to some drug companies. One of them was Glaxo, the very firm that not so long ago had tried to hire him away from Columbia. Glaxo now offered to support Nicholls to the tune of $80,000, spread out over eighteen months, in return for nonexclusive rights to use (but not own) the final product. That was about half of what he would have been making if he had taken the job at Glaxo, but it no longer made any difference to Nicholls. He readily accepted the deal.

Dave Weininger's forecasts seemed to have been entirely correct. Nicholls was happier being in business for himself in Santa Fe than he would have been working in Glaxo's lab, and he was far happier still than he would have been had he remained at Columbia.

"At Columbia I would have been focused on money: on getting grants, on the commercialization of the software," he said much later. "Paradoxically, it was when I broke away from the university and struck out on my own in Santa Fe that I really began to pursue science for its own sake."

EARLY IN 1995, about six months prior to writing his "You should do it" e-mail to Anthony Nicholls, Dave Weininger took his first ride in a military jet. Fantasy Fighters, a flying service at Santa Fe airport where Dave hangared *Puer Æternus*, his piston-engine Aircoupe, kept a small selection of the world's more famous military jets stationed on the field. These included a Mikoyan MIG-15, a Lockheed T-33 Shooting Star, and the Russian L-29 Delfin. If you were already a seasoned private pilot with a multiengine rating and a thousand hours or more of solo time (which Dave Weininger certainly had), then Fantasy Fighters would take you through a training course designed to get you the necessary legal clearances to operate any of their exotic combat aircraft. In the United States, the relevant document was known as an LOA—a Letter of Authorization—which was a statement from the Federal Aviation Administration certifying that you had demonstrated the ability to act as pilot-in-command for the particular type of aircraft in question.

Of the used military jets that Fantasy Fighters kept in its inventory, the L-29 Delfin was their entry-level model, the easiest of them all to fly, requiring only some five to ten hours of advanced instruction to bring the pilot up to the necessary level of competence. The L-29 Delfin was a single-engine, straight-wing craft with tandem seating and excellent visibility. The plane had been a standard jet trainer in Soviet Bloc countries for more than twenty years; some thirty-five hundred units of the craft had been produced, and many of them were still in active service, primarily in Third World air forces. The Delfin was rather slow as military jets went, with a maximum speed in level flight of 340 knots (about 390 mph), but had an entirely respectable rate of climb, three thousand feet per minute, which made for a thrilling ride, especially for a first-time passenger.

When Dave Weininger took his flight in the L-29, he was totally floored by the experience. For a while afterward, practically all he could say about it was, "Wow!"

He decided that he had to have one of his own—if not an L-29, then at least a jet of some sort. Flying a jet versus a piston aircraft such as the Aircoupe was the difference between driving a Volkswagen Bug through city traffic and putting a Ferrari Testarossa through its paces at Le Mans. By this time, Dave had become a competition-level stunt pilot and had won ribbons in several aerobatic events. He was a certified flight instructor. He had an instrument rating, a multiengine rating, and had logged time in helicopters. Still, all of that was immediately put into the shade by a jet, and with the success of Daylight as a company, he now had the wherewithal to own such a craft himself, personally.

Sales of used military jets constituted a little-known and shadowy backwater of an already narrow market, but an active pilot such as Dave knew the appropriate sources: *Trade-A-Plane*, *Flying*, and other aviation magazines ran ads for some of the wildest and weirdest airplanes ever produced, in addition to which there was also the Internet. From all these reservoirs of information Dave learned that it was possible to get a used jet fighter in reasonable condition for not much money at all—in the general neighborhood of a couple of hundred thousand dollars, which was no longer out of the question for Dave Weininger.

But then at about the same time as he was making his first flight in the L-29, a second opportunity fell into his lap. The Santa Fe house that Roger Zelazny, the science fiction and fantasy writer, had owned and lived in for twenty years had been put up for sale. On June 14, 1995, Zelazny died of cancer at St. Vincent Hospital at the age of fifty-eight. He had two sons and a daughter, but at the time of his death he was estranged from his wife, Judy, and was living with an English professor and author by the name of Jane Lindskold.

Dave Weininger was a science fiction addict and a fan of Roger Zelazny's, so when he heard that the writer's house was on the market, he had to go and have a look. The house proved to be a large and rambling structure on Stagecoach Road, which ran along a ridge north of the city, and when he and Dawn walked through the place, they found that it had not been kept up and was not in good repair. This was no obstacle to them:

having already remodeled a river barge, renovating a mere house would be child's play to Dawn Abriel.

And so they decided they had to have it. Dave was an amateur astronomer, and the house's ridgetop setting would make for excellent viewing in all directions. He even thought of putting up an astronomical observatory on the site, a structure with a rotating dome and moveable aperture, just like Mount Palomar. Beyond that there was the considerable symbolic and inspirational value of owning and living in the very house where one of Dave's literary heroes had done much of his work.

In the late summer of 1995, Dave and Dawn bought the place and moved in.

A few months later, Dave heard that a British jet trainer known as a T5A "Jet Provost" was for sale in rural Georgia. The seller, a professional arms dealer, was asking for $150,000, cash on delivery, at Griffin-Spalding airport. Located near the town of Griffin, about thirty miles south of Atlanta, the place billed itself "The Best Little Airport in Georgia."

The T5A was a fighter-bomber training aircraft, and the seller was offering all kinds of operational battle gear to go along with the plane itself. For only a slight amount of additional cash, the seller explained, he could outfit the plane with a set of matching cannons, bomb racks, or tip tanks, take your pick. On top of which he could also hook Dave up with a sister company that would supply him with "expendables"—cannon shells, bombs, rockets, whatever.

"Depending on quantity, they'll need a two-week notice," the seller advised.

All of this was news to Dave. He was not about to make any low-level bombing runs over the New Mexico desert or anyplace else, but here he was being offered a complete line of military ordnance as if it were no more than a styling package on a new car. He declined the offer of guns and ammo but decided to buy the plane itself.

In February 1996, therefore, Weininger arrived at Griffin-Spalding with a suitcase that contained approximately $150,000 in folding bills, handed it over, and took delivery of the plane. He was now the proud possessor of the single most exclusive item in private aviation, a military jet, a craft with an engine that emitted not a low roar but a high whine, a fighter-bomber that he could call his own.

RIGHT ON SCHEDULE, an enfilade of company officers assembled at the BiosGroup's premises in Santa Fe and patiently awaited their turn to see Stu Kauffman, doctor to ailing corporations. One thing the BiosGroup had going for it from the beginning was that its partner in crime, Ernst & Young, could funnel their own clients to Bios, or at least those who were experiencing problems. There was Southwest Airlines, for example, who needed help with its air freight operations.

Southwest, as it turned out, did not move freight as efficiently as it moved people. All too often, cargo arrived late at its intended destination, and much of it got misrouted, delayed, or lost as pieces were transferred repeatedly from plane to plane. Additionally, backlogs of parcels heaped up overnight in huge masses, causing flight delays, aggravating passengers, and generally making a mess of air operations. Despite all this, the company's cargo business was actually on the rise, and in fact it was rising at an even faster rate than was its passenger revenue, and the airline did not want to miss out on a cent of this much-needed source of income.

So Stuart Kauffman, M.D., now immersed himself in the intoxicating scientific problem of air freight. No matter. This was a man who understood *everything*; air cargo operations could hardly present any significant challenge to his talents. It was just another problem, of that he was sure, one that like all the others would be solvable by the application of empirical methods, rational analysis, and appropriate computer simulations.

First off, data gathering. Kauffman and his crew collected flight schedules, cargo manifests, bin utilization reports, and so on.

The primary question to be answered on the basis of this data was, what really happened when the airline's freight house received a package for shipment? The possibilities were as follows: (1) Ship it out immediately on the first departing flight that was headed in the right direction, even if that flight did not land at the package's destination city. (2) Hold the package in storage for the first flight that did serve the destination city. (3) Follow whatever routing was listed on the freight manifest, even if that meant holding the package in a storage bay for a considerable length of time. (4) Do some unknown but ultimately discoverable combination of the above.

The Bios scientists constructed computer models of each strategy, as well as permutations of them, and ran the simulations hundreds, indeed thousands of times, in an attempt to duplicate the airline's actual experience as reflected in the data they had collected. What they discovered from all this was that the airline's cargo handlers were actually following the first strategy, which was to say that they got rid of any package just as soon as they possibly could, after which it was no longer their problem.

The Bios researchers dubbed this "the Hot Potato rule." Considered abstractly, the Hot Potato rule made a certain amount of horse sense: the package was shipped as soon as it was received, whereupon it was in motion toward its destination. Once the plane containing the package landed, the package would again be put on the first flight that would take it toward, even if not to, its intended terminus, and this process would be repeated again and again, seriatim, until the parcel finally got to wherever it was addressed.

The Achilles' heel of the Hot Potato rule, however, was that the same package got handled multiple times, increasing the chances of misroutings, mishandlings, and delays at every point during transfers from plane to plane, and creating the dreaded bottlenecks where freight piled up in storage bins while waiting for the next flight out. All those transfers, moreover, made for extra labor costs and represented inefficiencies and wasted time for cargo handlers. Bad as the practice was with any one package, applying the Hot Potato rule to each and every item only compounded these problems exponentially.

Additional simulations revealed that by following an alternative rule, called "New Strategy," all those problems could be avoided. The New Strategy was to keep the package on the same plane, even if that plane did not fly in the right direction at first. It could fly quite a circuitous route, the cargo racking up some extra air miles in the process, but if it finally got to where it was going within the delivery deadline, who could complain? The chief advantages of New Strategy were that it minimized transfers, mishandlings, and overnight storage pileups, while it made for big savings in labor costs.

The Bios scientists formalized this in a piece of software called MANI-FESTER, which required putting freight on the first flight that landed at the destination city within the delivery deadline, no matter where it went in the interim. The simulations showed that by following the MANI-FESTER system, Southwest would reduce freight transfers by some 75 percent and cut freight labor costs by 20 percent. There might be minor reductions in on-time performance as the cargo traveled hither and yon around the country, but that drawback was more than offset by the savings in labor costs and by reductions in misroutings and overnight storage holdups. Overall, MANIFESTER represented a signal and clear advance over Hot Potato.

The Bios programmers also incorporated into the program a suite of features that would show the airline how to cope with sudden increases in freight volume, to add new routes into the system, and to estimate how freight revenues might be affected by changes in pricing.

The airline's officials listened to all this, accepted the software, and said they would give it a try.

AN UNPREJUDICED OBSERVER might wonder whether there was anything specifically complex systems–dependent in the BiosGroup's approach to the Southwest Airlines cargo-handling problem. Was this really anything more than common sense, aided and abetted by spiffy simulations? Indeed, on a superficial view of the matter, the entire situation represented an exception to rather than an illustration of the primary dogma of complexity theory.

If there was any precept that constituted the defining characteristic of complexity science, it was the doctrine that local interactions among independently acting agents frequently gave rise to emergent patterns of behavior not intended by any one of them. The development of orderly patterns in Kauffman's randomly programmed blinking-light experiments was a classic example of that phenomenon.

Often enough, the behavior that sprang forth turned out to be beneficial to the agents involved, and to the entire system of which they were a part—the "invisible hand" effect in a market economy, for example, where a number of private parties, each of whom acted strictly for their own selfish purposes, nevertheless tended to promote the common good. ("Every individual intends only his own gain," Adam Smith had written. "And he is in this, as in so many other cases, led by an invisible hand to promote an end which was no part of his intention.") There were plenty of other examples of emergent behaviors that were likewise useful to the agents, individually and collectively: the flocking of birds, the schooling of fish, the swarming of bees, illustrating if nothing else the value of safety in

numbers. Even the "V" formations of flying geese had been explained by
scientists as being artifacts of individual geese adjusting their flight paths
to take aerodynamic advantage of the wingtip vortices generated by the
geese ahead of them.

But in the case of the Southwest Airlines cargo handlers, the behavior
patterns that had emerged spontaneously from their individual actions
had been positively destructive, and their operations could be characterized fairly as yielding a complex maladaptive system rather than an adaptive one. The Hot Potato rule was part of the problem, not part of the
solution. The solution, in fact, ran entirely counter to Stu Kauffman's own
oft-stated preferences for bottom-up methods of resolving conflicts, for
decentralization, distribution of authority, and local control. The Bios-
Group's own solution, the so-called New Strategy, would operate in the
classic hierarchical, authoritarian, top-down manner, imposed from above
by management, no "local control" here. As Kauffman later admitted, in
the case of Southwest Airlines, "The magic that we worked was just to be
on the spot and to be reasonably smart about it."

Still, the empirical fact of the matter was that when the New Strategy
was implemented, it worked as predicted, first in short-term field trials
and again when it was officially implemented as company policy. During
the first full year in which Southwest employed the New Strategy, the
company reported an annual savings in labor costs of $10 million.

Southwest Airlines was only a beginning, and complexity science was
still young and unformed and by no means set in stone.

"Complexity science is fifteen years old, and attempts to apply it to the
real world are three years old: it's as old as Bios," Kauffman said. "The science is a teen-ager, and the attempts to apply it are just past the toddler
stage."

So nobody should get rattled if it didn't work immediate miracles when
first applied to the business world. Maybe it would work, maybe it wouldn't;
you simply had to try it out and see what happened.

The fact was that the BiosGroup had many other worlds to conquer—
they would end up advising clients that included Ford, Boeing, Texas
Instruments, NASDAQ, Honda, and Unilever, among others—and they
would utilize extremely advanced science and technology (in the form of
computer simulations), even if not complexity theory per se in some cases,
in an attempt to solve their respective problems.

There was the case of the Walt Disney Company, for example, owner and operator of the famed theme parks. Here was a business that started from nothing if there ever was one. "I only hope that we don't lose sight of one thing," Walt Disney himself used to say. "That it was all started by a mouse."

The various Disneylands and Disney Worlds had by now become so popular, however, that they were being victimized by their own success. As a detachment of company officers told the BiosGroup, visitors to their theme parks habitually clumped together at certain attractions at the very same times every morning, afternoon, and night. Regular as the tides, park visitors would show up at the rides and the restaurants, the shows and the shops, causing long waits and sowing dissatisfaction among customers. (Unfortunately for complexity theory, this was another example of bottom-up, spontaneous, emergent behaviors *causing* problems rather than solving them.)

The Disney managers decided that by eliminating or reducing these bottlenecks, they could increase not only revenue but also customer satisfaction by shortening the lines and waiting times and decreasing the overloads and congestion. The question was, how to get rid of the bottlenecks?

The Bios scientists responded by creating ResortScape, a piece of software which they described as "an agent-based model of the park that provides an integrated picture of the environment and all the interacting elements within the resort." Rides were represented by red blocks, shops by blue blocks, and restaurants by green blocks, while visitors were reduced to small circles. Running the ResortScape simulation was an exercise in monitoring "guest flow," which meant following the course of the tiny circles as they clumped together at the park entrance gates in the morning rush hour, dispersed out toward the various rides, and then clumped together again at food concessions and restaurants at lunchtime. Afterward it was back to the rides, to the souvenir shops, and then a hasty retreat to the exit gates and parking lots at the end of the day.

By trying out alternative scenarios, adjusting ride capacities to peaks in demand, adding special events, shows, or other diversions to take the load off the rides, extending the park's operating hours, and so on, the scientists were able to lessen the severity of the clumping phenomenon. And so in due course the Bios theme-park engineers had successfully figured out how to get more people to more rides in less time while reducing line lengths and waiting times and smoothing out guest flow in other areas of the park. It seemed like a perfect solution.

Unfortunately, the Disney managers never actually implemented ResortScape. They regarded the program "as basically 'eye candy,' " according to a high-ranking officer at the BiosGroup. "It was well received by Disney but never made it into implementation after the project sponsor departed the company."

SCIENCE WAS INHERENTLY neither a moneymaking nor a money-losing activity, although it was widely considered somehow "purer" to lose money than to make it. In the fall of 1980, when Harvard University was faced with the decision whether or not to become a minority shareholder in a biotechnology company in which two of its faculty members would play a prominent role, the university decided not to participate, saying that to do so would threaten "the preservation of academic values." In announcing Harvard's conclusion, President Derek Bok cited as primary two reasons why academic scientists should beware of making money with their science.

"First, the prospect of reaping financial rewards may subtly influence professors in choosing which problems they wish to investigate. Academic scientists have always feared what Vannevar Bush once termed 'the perverse law governing research,' that 'applied research invariably drives out pure.' . . .

"The final danger is a threat to the quality of leadership and ultimately to the state of morale within the scientific enterprise. The traditional ideal of science was based on a disinterested search for knowledge without ulterior motives of any kind."

That may have been the traditional ideal once upon a time, but by the end of the twentieth century it was an open question whether scientists, or society at large, could afford any longer the "disinterested search for knowledge" to the exclusion of advances in technology, applied science, and the pursuit of scientific knowledge for the sake of financial gain. By the mid-1980s, indeed, the financial cost of doing certain types of "pure" scientific research had far outstripped the ability or willingness of society to pay for it. There was the case of the Superconducting Supercollider, for example.

Proposed by a small cadre of American particle physicists in the early 1980s, the Superconducting Supercollider, or SSC, was to be the last word in particle accelerators. If and when it was ever constructed, the device would be the largest machine on the face of the earth.

Other than the earliest experimental models, some of which could fit on a lab bench, particle accelerators were big by nature. Fermilab, built in the 1960s on open farmland outside of Chicago, encompassed more than a hundred buildings and employed fifteen hundred workers who collectively operated a million tons of heavy equipment. The centerpiece of Fermilab was "the main ring," the accelerator proper, which was an enormous circular loop of magnets buried under a mound of earth. The ring was four miles in circumference, a physical feature so large and distinct that it was clearly visible in satellite photographs taken from three hundred miles away in space, showing up on the midwestern landscape as a thin and perfect circle.

Big as it was, Fermilab was tiny compared to the SSC. The SSC, according to the central design group's final specifications, issued in April 1986, called for a ten-foot-wide underground tunnel forming an oval ring that would be fifty-three miles in circumference—as big around as the Washington, D.C., Beltway, and large enough to enclose almost the whole of Rhode Island. The SSC would not be made of anything so cheap as concrete or asphalt, however. Rather, it would consist of 4,728 magnets, each of them 50 feet long, containing altogether 41,500 tons of iron and 12,000 miles of superconducting cable, the whole colossal arrangement kept within operating temperature limits by 2 million liters of supercooled liquid helium. The price tag for all this was to be $8 billion—as if initial estimates were ever correct. And that was only for building expenses; operating expenses would require additional large sums, to be provided on an annual basis.

The hope was that with this machine, physicists would discover the elusive "Higgs boson," an elementary particle that was forecast by theory but never yet seen in any accelerator so far constructed. Of course, the SSC might (or might not) also discover other things. Texas physicist Steven Weinberg, one of the most fervid champions of the SSC, described these possible findings: "Particles within the quarks that are within the protons; any of the various superpartners of known particles called for by supersymmetry theories; new kinds of force related to new internal symmetries; and so on."

"We do not know which if any of these things exist," he added, "or whether, if they exist, they can be discovered at the Supercollider."

This entire cosmic superappliance, in other words, could conceivably fail to uncover any of the entities it was built to produce and detect. Nor would this be entirely surprising: its purpose, after all, was to perform experiments, and experiments were known to fail. But even if they succeeded, the identity, nature, and significance of the new particles would be understandable only to an exceedingly small portion of the world's population—if, indeed, even to them.

Despite the somewhat vaporous nature of the "pure" knowledge to be furnished by the device, many U.S. senators were initially in favor of building it—as long as it would be located in their home states. As one of them told Weinberg during committee hearings in 1987, almost a hundred senators were in favor of the SSC, but after the site was announced there would be just two. House Representative Don Ritter, for that reason, called the SSC a "quark-barrel" project. Finally deciding that the country could not afford this particular piece of lunatic-fringe scientific megalomania, Congress quietly killed it.

The SSC epitomized "big science," pure science, and pure scientists as vast money sinks, or at least as net consumers of wealth. The fact was, however, that applied science, even when pursued for financial gain, could produce equally great, if not greater, payoffs when it came to the advancement of knowledge—even pure scientific knowledge. Consider, for example, the case of Leroy Hood and his "gene machines."

Hood was a Caltech biologist who in the mid-1980s had an idea for building a type of machine that would be useful to scientists pursuing basic research in molecular biology. The device was a DNA sequencer, and its purpose was to read the order of complementary base-pair sequences on strands of DNA. The machine would work by taking strands of DNA and splitting the molecules down the middle. Then, by performing a series of chemical tests, the machine would determine the exact ordering of the exposed bases, which it would then print out on a list, for example, AACGTGGTC.

Such a device, if it could be made to work reliably, would allow scientists to automate the process of DNA sequencing so that they could read out the genomes of entire biological organisms, something then considered as still largely in the pipe-dream stage. The machine could also speed up the process of DNA fingerprinting in criminal cases. It could also make the inventor lots of money.

Hood and his colleagues at Caltech had previously invented a machine that performed essentially the same task with proteins: it broke down protein molecules and identified each of their component amino acids in turn until it had determined the entire linear succession of them. Hood had been convinced that a protein sequencer would be useful to researchers studying protein function, by determining how the molecules operated and how they contributed to health and disease (Hood had a medical degree from Johns Hopkins). And so he visited a succession of high-tech instrumentation and chemical companies in an attempt to commercialize the invention. He visited DuPont, Beckman Instruments, and so forth, explaining what a boon these protein-sequencing machines would be to molecular biology, the drug industry, and the progress of medicine, if they would only consent to manufacture the things. In all, he called on nineteen different companies, and every last one of them declined the honor.

"They're nice machines, but nobody really needs them," he was told time and again. "We just wouldn't sell that many units. There's no money to be made with this."

So in 1981 Hood and a group of partners, with backing from a San Francisco venture capitalist, founded their own company, Applied Biosystems, in Foster City, California, to manufacture and market the protein-sequencing devices. They proved to be highly popular and commercially successful, and advanced the state-of-the-art understanding of the structure and function of protein molecules.

Five years later, Applied Biosystems merged with PerkinElmer and brought out Hood's new DNA machine, the Model 373 DNA Sequencer. Within a short time, the Model 373 DNA Sequencer became one of the most successful analytical instruments in the world, racking up sales of more than three thousand units in ten years, at the average price of approximately $110,000 per machine.

Hood and his partners had become multimillionaires. In the process, they had provided employment to thousands of workers, created huge amounts of new wealth for investors, and contributed to the scientific knowledge of the human genome. It was Leroy Hood's DNA sequencers, in fact, that had made the Human Genome Project possible.

"The genome revolution would not have happened without them," said Craig Venter, head of the privately funded genome-sequencing firm, Celera.

In Santa Fe at the beginning of the new century, the Info Mesa's scientists were squarely in the Leroy Hood mold. Neither Stu Kauffman, Dave Weininger, Anthony Rippo, Anthony Nicholls, nor any of their peers had qualms about doing science for profit, about the "corrupting" influence of money, or about the specter of "applied research invariably driving out pure research."

To the contrary, they viewed the prospect of using science to make money, as opposed to using money to do science, as an essentially noble goal. The moneymaking scientist, after all, provided goods and services to clients, created new jobs and new wealth, increased the sum total of scientific knowledge, and by doing all this, helped people.

"We want to make drugs that make people's lives better," said Anthony Nicholls of OpenEye Scientific Software. "We're not going to shave 1 percent off the cost of making a car for GM—we're going to make something that actually affects people's lives. If in the entire arc of OpenEye, all we ever do is speed up the discovery of one drug by six months, it would be worth it."

BIOREASON SOON OUTGREW its Johnson Street offices and in 1999 moved to the Wells-Fargo Building in downtown Santa Fe, a modern adobe structure a block off the Plaza and across the street from the city's most upscale and expensive hotel, the Inn of the Anasazi. The firm occupied most of the third floor, around which there were balconies that offered fine vistas to the north and south. Anthony Rippo moved into an outside office, and every once in a while he would promenade out onto the balcony through the French window-door and play a few experimental measures of Vivaldi or Tartini on one or another of his homemade violins or violas. All in all, it was an exceptionally choice and comfortable spot, and the company would remain there for some years to come.

In its initial real-world trial-by-fire, the company's drug-discovery software module had performed somewhat better than expected. Going into the experiment, Tony Rippo had hoped that Bioreason's software would discover in the Parke-Davis data screen at least some of what the drug company's own medicinal chemists had found in it.

"I'm thinking, I hope we find most of everything they found," Rippo said later. They would look like idiots if the software found nothing in the data set or discovered relatively little. But in fact the software beat Parke-Davis's own drug-discovery team on its home ground.

"In hours of computing time, not only did we find everything the Parke-Davis scientists found, we found three other things they had not found," said Rippo. The software had identified some false positives in the Parke-Davis data, some false negatives, and, the hardest things of all to find, a few "outliers" or "singletons," compounds that had potentially druglike properties but were off in some bizarre chemical corner by themselves, not in any broader molecular family.

The software had done all this by taking the incoming data and organizing it, recognizing chemical themes based on common molecular scaffolds, and then classifying the compounds into families, data trees that exhibited genealogical relationships among the chemicals. At the end it presented the results to the user, showing the common elements in red, the inactive compounds in yellow, and the promising active compounds— the next big blockbuster drug lurking among them, perhaps—in green.

The final payoff was that Bioreason's automated chemical reasoning system had done better, in a matter of hours, than Parke-Davis's own in-house medicinal chemists had done in six years.

"That," said Rippo, "was a blowaway."

To make sure that it was not a fluke, the luckiest computer quirk in history, Parke-Davis then sent Rippo and the others a second data set, one that they thought contained within it potentially even more important chemical compounds than the first. Bioreason's software system acquitted itself equally well in this case too, spotting a further range of false positives that Parke-Davis scientists hadn't identified as such. Looking at them again, the drug company's chemists acknowledged that the software was correct, and that its findings would spare them a major amount of time and trouble they would otherwise have spent following up on false leads.

Bioreason's software had discovered the gold nuggets that lurked within the data sets, and had separated them from the fool's gold. Plainly, their automated reasoning systems were more than eye candy.

11

WINNING THROUGH FENG SHUI

IN 1997, DAYLIGHT introduced a new software module called Reaction Toolkit, a program with which a chemist could in fact do "virtual chemistry." As Dave explained it, "A chemist using this system can do a million experiments in one day. If he's unsatisfied with the result he can do a million experiments the next day, and then go back and hit the three of them that look promising, then actually do the wet chemistry on Wednesday, and have it written up that weekend."

Reaction Toolkit was not Daylight's only new application, however. By that point, the company had incorporated into their growing list of product offerings practically every last chemical database they could find, no matter what the source. They had added to their collection the *World Drug Index*, a compilation of sixty thousand drugs and pharmacologically active compounds, including all of the world's marketed drugs; *Spresi*, a list of 3.2 million substances compiled by Russian and German researchers; *Current Chemical Reactions*, a list of some 250,000 separate types of known chemical reaction patterns; *Index Chemicus*, a listing of the chemical structures that had been added to the world's compounds since 1993; the *TSCA93 Database*, a catalogue of the more than 100,000 substances covered by the U.S. Toxic Substance Control Act; and so on and so forth—more than a dozen separate databases and about two-dozen distinct applications and toolkits. All of these modules had been brought into Daylight's burgeoning system software so that each worked smoothly and seamlessly with all the others, the new releases integrated with all of the various add-ons, patches, revisions, and refinements that the company had made to its systems since its previous software update. And all of the new modules, of course, were made fully compatible with and accessible by the Thor and Merlin search engines.

With these numerous applications, toolkits, and sources of chemical information now centralized at Daylight, it was clear that the company was becoming a world focal point for chemistry, a fact that was reflected in the firm's overall sales and revenues. Dave and Yosi had become multimillionaires. That didn't stop them from continuing to barnstorm the country, hawking their wares as if they were still in the startup stage. When a relevant scientific conference or trade show was in the offing, they'd stash a few workstations in the trunks of their cars and lug them to the site. They'd set out their literature displays, fire up their machines, and sit, waiting for interested parties to stop by. After all this time, they were still having fun!

"There is a big fun factor for me at trade shows," said Yosi Taitz. "People come up and I tell them 'I can search a million molecules in under a second.' The person would say, 'Yeah, right.' And then I do it. The person would think this was a prestaged demo, but in fact we were doing the search in real time."

Once in a while the demo results surprised even Dave, who liked to tell the story of the time he had one up and running at an American Chemical Society meeting when a guy came up to him at the display table and asked, "Are there any Japanese patents for Best?"

Weininger, who had no idea what this was all about, said, "Best?"

He typed in the word "Best," pressed <ENTER>, and in about a second, no more than that, Merlin returned the news that "Best" was the trade name in Argentina for a compound known as diazepam (Valium in the United States), for which a Japanese patent had indeed been issued. The search also unearthed a color picture of the compound's molecular structure, listed its chemical reactivity sorted by several different parameters selectable by the user, and showed the drug's possible synthesis pathways, reaction products when combined with other drugs, along with much other chemical data, arcana, and lore.

Its range of products having become generally accepted, purchased, and regularly used by mainstream drug, chemical, and agricultural companies, Daylight was making lots of money (Yosi, talkative about every other subject, would never say exactly how much). And because the company had virtually no overhead—it had no large physical plant, maintained no storage warehouses, and had no standing inventory, transmitting its systems directly over the Internet—it was fast becoming, on a per capita basis, one of the more profitable small companies in the United States. Still, by 1997 Daylight had about a dozen employees working in the single small building (the former shelter for unwed mothers) on East Palace Avenue, and they were running out of elbow room. Dave and the others looked at buildings all over Santa Fe but, finding none that they liked, decided to build one of their own.

There were a few vacant acres on Route 285 north of the city, adjacent to the Santa Fe Radisson. The hotel had been put there for a reason, for it was a highly photogenic, almost romantic spot. This was clearly the place to build the new headquarters, the price was right, and so they bought the property, cash on the barrel.

Construction year for the new Daylight building was 1999, and several other Info Mesans attended the official groundbreaking, including a full detachment from Bioreason.

"John Elling and I rented a backhoe and had it delivered to the parking lot over the hill from the building site," Susan Bassett recalled. "All the Bioreasoners then climbed up on the backhoe. We rode it over to the groundbreaking ceremony and unfurled a banner that said, 'We're Here To Help.' It's one of the few times I've ever seen Dave Weininger actually speechless."

When it was finished, the new Daylight facility proved to be a three-story L-shaped building set artfully into the hillside. On the top floor, Dave Weininger's office looked out across the city and beyond it, toward the Sangre de Cristos. The view from his office was particularly striking in late evening as the sun sank toward the horizon and cast its dying orange rays on the already reddish mountain cliffs. Not that Dave was ever there to witness this effect: other than for unavoidable company meetings, he still did most of his work at home.

More of a challenge than putting up the building itself was the problem of getting city permission for the molecular sculpture that Dave wanted to erect in front of the place. The sculpture was to be a large, towering composition, a ball-and-stick model of the atomic structure of a cognition-enhancing drug, one that never made it past the experimental stage. Technically, the chemical was called quinuclidine-3-spiro-5'-methyl-2'-tetrahydrofuran-4'-spiro-1,3-dithoilane. Even the SMILES for it was simpler:

CC1OC2(CC31SCCS3)CN4CCC2CC4

Dave imagined that this molecule would symbolize the company and what it was all about—and besides, to him the object was highly pleasing on an aesthetic level.

The Santa Fe city government, however, did not agree with any of this. To them the structure represented advertising for Daylight, and they were not about to issue a building permit for such a crass and vulgar display, especially at a locale where it would command so much attention.

But the object was not "advertising," Weininger replied, it was a *sculpture*. For one thing, the molecule was not the company logo, which was, instead, an image of the sunrise. For another, the company was not in the business of creating, storing, or selling molecules, whether of cognition-enhancing drugs or anything else. They were selling "chemoinformatics tools"—information, software systems, databases, languages—*spiritual* things, in essence, not "objects." This appeal to spiritual values was more than satisfactory to the city fathers—this was Santa Fe, after all—and they duly let him erect the thing.

Dave actually built the object himself with the help of a metalworker friend of his (Dave always liked to acquire new skills), in return for which he gave the man flying lessons.

And so the object was built, and there it stood, a dozen or so color-coded, basketball-size steel spheres held together by beefy steel tubes, all of it overlooking Route 285 and the rest of Santa Fe, an accurate three-dimensional representation, Dave said, "of a highly rigid analog of acetylcholine, the first reported M1 selective muscarinic agonist," whatever that was.

IN ITS FIRST four years of operation, Bios had undertaken projects for about thirty different Fortune 500 companies in the United States, and in addition took on some classified work for the Joint Chiefs of Staff and the

U.S. Army Intelligence and Security Command. During that time the firm had acquired a staff of sixty employees, and by early 2000 had expanded its offices to occupy the entire first floor at 317 Paseo de Peralta.

Satisfying as all of this was to Stu Kauffman the scientist, what gave him even greater joy was confronting one of these corporate dilemmas and discovering a bit of new science in the process. This had happened in the case of Procter & Gamble.

Procter & Gamble was a complex system par excellence. It was a major global corporation, selling some 250 different products to nearly 5 billion consumers in more than 130 countries around the world. Indeed, there were few people on the planet who were untouched by its product line, which ranged from Crest toothpaste to Pampers disposable diapers, from Tide detergent to Oil of Olay. The company had 106,000 employees working in over eighty countries, a vast workforce that turned out a steady flow of merchandise, converting raw materials to a variety of finished products and then distributing them systematically to all corners of the earth.

In order to keep it all going, the company bought and took delivery of great quantities of raw materials, substances that it processed along an exceptional number of parallel and intersecting manufacturing pathways, transforming immense volumes of ingredients into a variety of trademark wares. It churned through an avalanche of natural resources on a daily basis.

But in 1998 Procter & Gamble came to Kauffman in hopes of "optimizing its supply chain." The fact was that its whole operation existed on so gigantic a scale that even a small improvement in the acquisition, utilization, distribution, and processing of raw materials and supplies could yield a significant decrease in operating costs, especially in view of the fact that all the relevant phenomena were continual and recurring. Progress in any one area could conceivably have "synergies" elsewhere in the system, as the latest corporate buzzword had it.

Anyway, a few members of Procter & Gamble's senior management arrived at Bios and asked for help in perfecting their entire "earth-to-earth" supply chain—their whole long trail of resource allocation, manufacturing, distribution, and final end-point use by the consumer. Even a 1 percent increase in their supply-chain efficiency, they said, could save them major sums of money.

Ironically, this was a problem that Procter & Gamble's own analysts couldn't address because in point of fact, they didn't know what their own supply chain was—not conceptually. They were responsible for it, they operated it, they oversaw it, ran it, used it, and paid for it, but as far as understanding it on any theoretical level was concerned, the company managers were pretty much at a loss.

So in their now standard fashion, Stu Kauffman and his team gathered data and made a full study of the Procter & Gamble supply chain. They found that it was characterized by three major parameters: total inventory in the system, total time in the system, and out-of-stocks on the shelves. Of these three "figures of merit," the only one that couldn't be modified was out-of-stocks: without exception, Procter & Gamble wanted to have Tide, Comet, and the rest of their products on the shelves at all times.

The Bios scientists ended up making five different simulations of how the Company's supply chain operated. They ran the simulations thousands of times under different settings and conditions, creating, in Kauffman's words, a "policy space with lots of knobs you can tune." As they observed the results, the scientists noticed that one particular effect kept cropping up: the sudden appearance of what Kauffman identified as "lumpy integer constraints."

A lumpy integer constraint, in this case, was the company's requirement that any given amount of input or output must be in terms of whole numbers. Procter & Gamble had introduced such a directive into its supply-chain operation by imposing a general rule on its cargo trucks that all shipments were to be made in full truckloads only; partial loads were not permitted. Such a strategy made obvious and intuitive sense: having all the trucks full when they departed the loading dock maximized their utility, left no wasted space, saved on diesel fuel, reduced air pollution, minimized duplication of effort, and had various other advantages.

But there was a hidden downside to all this. For what the Bios-Group's simulations found was that adherence to the full-trucks rule caused perturbations elsewhere in the system, as trucks waited around unproductively for their cargo holds to be filled when they otherwise could have gone out to restock store shelves with needed items. Kauffman and the Bios crew had discovered that the full-trucks rule converted smooth (or "laminar") flow into turbulent, irregular, or jagged flow, creating bottlenecks and even temporary out-of-stocks in the stores. Or as Kauffman himself expressed it, "These lumpy integer constraints created conflicting constraints that made the landscapes rugged as hell."

Fitness landscapes had finally made their appearance in a business setting. They had suddenly arisen out of the supply-chain gloom where they showed themselves to be deformed and irregular, and all on account of these unwanted lumpy integer constraints (the full-trucks rule).

Here, finally, was complexity science coming into its own within a strictly business context. Here was an apparently harmless but in fact subtly damaging top-down requirement causing hidden problems on lower levels within the system. The classic complexity-theory article of faith had at last been illustrated in practice.

The solution, of course, was as plain as day: Relax the full-trucks rule.

"We discovered," says Kauffman, "that if you softened the integer constraints just slightly—so that you could send less than full truckloads, sometimes—if you just smoothed them over a little bit, then the laminar flow regime was reestablished, you stabilized laminar flow.

"This is a scientific discovery," he added, "and we're 98 percent certain of this answer. If you soften the integer constraint, you can restabilize laminar flow."

As he had earlier patented his ideas for making proteins by the million, Kauffman now of course wanted to patent this new discovery as well.

"The patent's pending," he said. "We have patents pending on everything. You name it, we've patented it. If you name it and we haven't patented it, you'll be a coinventor, and we'll patent it."

Strange as it might have seemed for a discovery related to lumpy integer constraints, rugged fitness landscapes, chaotic regimes, and laminar flow to be submitted for patent, the fact was that far more surprising items than these had been patented by scientists. One of the original inventors of the atom bomb, Leo Szilard, who had been residing in London at the time, had submitted to the British patent office in March 1934 his idea for the transmutation of elements. Shortly thereafter, in July of the same year, Szilard submitted for patent protection the idea of a nuclear chain reaction itself—the very principle underlying the atomic bomb. You merely assembled a mass of radioactive materials, Szilard said in his patent application, and started neutrons coursing through the mass: "If the thickness is larger than the critical value, I can produce an explosion." Duly impressed, the British granted both applications, awarding them patent numbers 440,023 and 630,726, respectively. (The A-bomb patent had been kept secret for thirteen years and was not published until 1949.)

Szilard, however, represented an extreme case of a patent collector. Szilard and no less a figure than Albert Einstein had patented a new type of refrigerator. Later, together with Enrico Fermi, he received a patent on the first nuclear reactor. And in 1953, when Watson and Crick had discovered the molecular structure of DNA, Szilard suggested to James Watson that he patent the double helix. "But he knew that his proposal would not fly," said Watson, "and told me I could only be famous, not rich."

Compared to Szilard, Fermi, and some other scientists, Stu Kauffman's patent claims were relatively modest, and in the case of Procter & Gamble he merely wanted to make some inroads into the supply-chain market.

With his work for Procter & Gamble, and with patents pending on several, potentially revenue-producing scientific discoveries, Stu Kauffman had finally completed the transformation from academic biological researcher to successful businessman-scientist. The metamorphosis was reflected in the new corporate motto for BiosGroup: "Science for business."

BY THE YEAR 2000, a little more than three years after its founding, OpenEye Scientific Software had developed into a flowering and flourishing concern. Besides Anthony Nicholls, who had retitled himself "El Presidente and Founder," the firm then had five full-time employees, none of them on the graduate-assistant or even postdoc level: they were already accomplished professionals with a full string of achievements behind them.

There was Matthew Stahl, who had a Ph.D. in chemistry and whom Nicholls had managed to lure away from Vertex Pharmaceuticals of Boston, one of the first structure-based drug-design companies, where he had been manager of ChemInformatics. Stahl was OpenEye's first employee, other than Nicholls himself, and had come to the firm in the summer of 1998. At OpenEye, Stahl wrote a program called Omega, so named because it was the final structure-generation program he ever hoped to write. The application handled large chemical databases and combinatorial libraries and permitted exhaustive searches of druglike molecules in fractions of a second. When correctly configured, Omega could screen virtual combinatorial libraries at the rate of some fifty thousand compounds a day.

There was Joe Corkery, from Princeton and Harvard Medical School, who joined the company in October 2000. There was Geoff Skillman, M.D., from Stanford and UCSF. And there was Mark McGann, with a degree in chemical engineering from Tulane, who became OpenEye's vice president of communications.

Nicholls's greatest coup, however, was hiring away one of Dave Weininger's own employees, Roger Sayle. Sayle was the author of the world's most widely used molecular graphics program, RasMol, and had produced another application that allowed the user to see three-dimensional images of molecules on a two-dimensional computer screen. Two separate images of the molecule appeared on the screen, each shown from a slightly different perspective, just as in normal three-dimensional viewing. To get the object to appear in three dimensions, the viewer had to stare at a spot between the two images and then cross his or her eyes, and gradually the two images would coalesce into one. When they did, the chemical molecule seemed to exist in three dimensions and hang there suspended in space.

Collectively, OpenEye's "New Mexico Six" would create a growing list of products. ZAP, which at the company Web page was represented by a bolt of lightning, calculated molecular electrostatics forces. ROCS (Rapid Overlay of Chemical Structures) was a shape-based chemical search tool. It hunted through three-dimensional structures at the rate of three hundred to four hundred different shapes per second.

VIDA, Anthony Nicholls said, was "the OpenEye philosophy made perceptible." It was an all-purpose graphics program designed to visualize, manage, and manipulate large volumes of chemical information at extremely fast speeds. There were competing programs on the market that did similar things, but most of them had an upper limit on the number of structures it could work with, usually a thousand or so. VIDA, however, easily digested hundreds of thousands of structures at a clip. Prices ranged from $12,000 to $25,000, although Nicholls made some programs available to colleges, universities, and government agencies for free.

Nicholls was not yet a rich man, but by now he had acquired many of the normal accouterments of the successful businessman. These included new levels of stress consequent on the obligation to meet payrolls and to fund employee benefits, buy insurance, and pay office rent. In 2000, OpenEye moved into a suite of offices at 3600 Cerrillos Road, one of the main business routes of Santa Fe. The office was in a mixed commercial/residential complex called The Lofts, an airy modern structure with skylights, double-story interiors, sunrooms, balconies, and other amenities.

Initially Nicholls rented a smallish twelve-hundred-square-foot office (half of which was devoted to Ping-Pong), but OpenEye soon expanded to a more serious twenty-seven hundred square feet. The new place included a guest suite for visitors, two kitchens, two bathrooms, a shower, machine room, reading area, and business office. "We did finally get a photocopier," Nicholls said. "There is a coffee maker but it doesn't get used: the tea drinkers won!"

IN NOVEMBER 2001, Bioreason had its reception area "feng shui-ed."

The lobby had been a sore point in the mind of Susan Bassett, the company president, almost from the beginning. The room was the first thing you saw as you entered the office, and it had been painted in some sort of indefinable tan: a harvest-gold, pureed-baby-food combination of hues that Susan Bassett had grown to hate. One morning she came in, took one look at the walls, and said to the company controller, Kathleen Eckhart, "Look, this color scheme is not right. It's got to go. I can't stand it anymore."

Kathleen Eckhart was not averse to making the change. Along with countless other Santa Feans, however, she was a strong believer in feng shui and suggested that it would be a mistake to do any major redecorations without doing a complete feng shui analysis of the situation. She therefore proposed bringing in a professional feng shui consultant in order to get a grip on the problem.

Feng shui (pronounced *fungk sway*) was an ancient Chinese discipline of obscure origins that dated back some three thousand years, when it had been used to determine the most propitious sites for tombs. Later its principles had been extended to establishing the sites of royal palaces, government buildings, monuments, and the like, until finally whole cities were being designed and built according to the tenets of feng shui. The underlying presupposition of it all was that the earth's energy fields somehow affected human life, and that proper alignment with these fields and forces could influence one's personal fortunes for good or ill. (Many Chinese businessmen in Hong Kong let feng shui practically rule their business lives, especially in the matter of the siting, orientation, and interior design of office buildings.)

Feng shui was wildly popular in Santa Fe, many of whose residents would hardly make a move, let alone paint a room, without first viewing the proposed course of action through the lens of feng shui. To help in this effort, Kathleen Eckhart enlisted the advice of one Loralee Makela, feng shui consultant, of Santa Fe.

For a feng shui specialist, Loralee Makela had impressive credentials, starting with lavish testimonials from satisfied customers. "My first experience with feng shui was a short seminar where I learned that I had never slept in the correct direction," said one of her clients in a glowing endorsement. "I immediately changed the orientation of my bed and the bursitis I had in my shoulder went away!!"

Makela's roster of clients included several realtors and other small businesses, a dentist, and a women's gynecology practice. She had been featured on the local Channel 7 News and on NPR's "All Things Considered." She also had a corporate Web site (makelafengshui.com) that listed among her services "Personal feng shui," "Feng shui environmental analysis," and "Yearly feng shui adjustments."

Sue Bassett was not a native Santa Fean—she was the daughter of a north Florida tobacco farmer and had been trained as a biophysicist and computer scientist. She was the last person in the world to get mystical, superstitious, or paranormal about interior decoration or anything else. Nevertheless, for the sake of getting the reception area repainted, and to placate the company controller, she agreed to have Loralee Makela come in and give the place a going-over. Bassett volunteered to pay for it out of her own pocket—she was not about to use any stockholder money for this.

So Loralee Makela came in, did a feng shui analysis, and submitted her diagnosis.

"Basically I found that their business liaisons area was in the lobby, and chose colors and elements that would enhance that energy," she said later. "Enhancing the energy makes that aspect of their business stronger rather than weaker, as it was with the existing decor."

Then, over the 2001 Thanksgiving weekend, Kathleen Eckhart and some friends executed Makela's suggestions. They repainted the walls a deep rich terra-cotta. They repositioned the plants, hung up a pleasant selection of water-themed prints, and installed wind chimes.

"Wind chimes!" Sue Bassett recalled. "And they had to be a very specific kind of wind chimes: five-note, five-tube wind chimes."

Makela explained, "The wind chimes are of the metal element which enhances business as well as calming down any negative aspects that could hinder their business."

At the end of the overhaul, Ms. Makela came in and gave a little seminar to the assembled Bioreasoners, explaining the renovations so that all those present would understand the reasons for each separate change. The company scientists listened to all this with due respect in the properly accepting, tolerant Santa Fe manner, but they were mentally rolling their eyes. Nevertheless, even they had to admit that the new lobby was a visual knockout, done in fiery earth tones that seemed to herald the rising of the sun.

What a surprise to them all, then, that at the start of December, only a few days after the official grand reopening, Bioreason suddenly did a landslide business selling their automated reasoning systems for drug discovery.

"Five purchase orders came in the next week!" Susan Bassett said. "And we landed a major deal that we'd been working on for months. I'm serious!"

Winning through feng shui?

"Who knows?" said Bassett. "Maybe it was just businesses making end-of-year purchases. I don't know."

Loralee Makela, for her part, said, "The space was brought into balance and the energy began to flow. Feng shui is not about forcing something to happen. It is about removing blocks and allowing what wants to happen happen. They had been doing lots of good stuff and just needed a little jump!"

The jump may have been provided by Bioreason's elaborate new software release, ClassPharmer. As with many other software systems, this one had been a long time in development. After their shakedown, "blowaway" tests of their prototype "E-Ruth" software, Bassett, Nutt, and the other Bioreasoners had set out to perfect a matched set of automated reasoning systems for drug discovery. At that time the global pharmaceutical industry was a $200-billion-a-year business, which as a whole spent some $56 billion on the research and development of new drugs. Americans, meanwhile, were using billions of pills of some forty thousand different types annually.

Those numbers, grand as they were, gave a wholly misleading impression of what it was like to be in the drug business. From the outside, it appeared to be a simple matter of finding some likely drug substances, stamping out pills, and then charging exceptionally high prices for them, raking in easy profits. The reality was somewhat different, for the process of coming up with a successful new drug was one of the higher-risk business activities. The fact was that for every 100,000 compounds that were discovered, created, or imagined by drug companies in pursuit of new products, only about 10 percent of them, or 10,000, showed even the slightest promise as drugs. Out of those 10,000, only a further 10 percent, or 1,000, went on to clinical trials for the purpose of establishing whether they were safe, effective, and worthy of being developed further. And out of those 1,000, only 1 percent, or 10 compounds, were ultimately brought to market and sold as new medicines.

Even then, not all of those ten compounds made a profit for the company that developed them. For every ten drugs that finally made it to the marketplace, only three even recovered their research and development costs, while only two of those three returned a profit to the company that had invented, developed, and tested the drug.

Moreover, in an age in which many American companies were criticized for making a fast buck and having short-sighted planning horizons, the drug business was characterized by some exceptionally long-term development periods, for the process of winnowing the good from the bad compounds was not normally a rapid one. Gleevec, touted by the media in May 2001 as an anticancer wonder drug that sprang up virtually overnight, had been discovered eight years earlier, in 1993, by Novartis (then called Ciba-Geigy), where it was known as a nondescript "compound STI-571." But the original research that led to the compound had gone back nearly forty years, to the late 1960s.

Nor was the drug development process cheap: the average cost of discovering a new compound, putting it through clinical trials, and bringing it to market was on the order of $350 million per drug, a figure that did not include the cost of the facilities and equipment needed to manufacture the substance in quantity. What all this meant was that in order to finance further research, pharmaceutical companies had to bring in three "new chemical entities" (NCEs) to the marketplace each year. During the late 1990s, however, the industry track record stood at 1.1 NCEs per company per year. To keep themselves afloat then, drug firms would have to triple the rate at which they converted NCEs into revenue-producing products.

Unfortunately, the need for speeding up the NCE revenue stream came at just the time when it was becoming progressively more difficult for any drug to be medically successful. The reason for this lay in the fact that historically it was the "easy" diseases that had been cured first, those that had single, well-defined, well-understood microbial causes (such as syphilis, for example). These were the conditions that had been vanquished by the first wave of wonder substances: sulfa drugs, penicillin, and other antibiotics. Some of the principal diseases awaiting cures at the start of the twenty-first century, by contrast, were ones whose causes lay within the body's cells (autoimmune diseases) or had multiple causes, some of which changed over time (such as AIDS), or whose causes were not fully understood (such as Alzheimer's).

For all these reasons, drug companies were willing to entertain virtually any method, product, or technology that seemed to promise a shorter, faster, or more streamlined path from discovery to market. Primary among them at the turn of the century was the practice of high-throughput screening (HTS), whereby robotic systems could create and test a thousand or more compounds a day, and later, ultra-high throughput screening (UHTS), which was expected to be producing up to 100,000 compounds a day when perfected. (In 2002, Vertex Pharmaceuticals claimed to be screening a record 2 billion "virtual molecules" a day, but these were imaginary compounds, not real ones.) The problem was to make sense out of the findings produced by such vast chemical data screens. Even a method that "failed" drug candidates earlier in the game, eliminating them from the research and development process before large sums of money could be spent on additional development, would repay substantial benefits in savings.

This was precisely where Bioreason's new product, ClassPharmer, entered the picture. Since there was no reliable way of going from a high-throughput data screen directly to an individual drug molecule, what the medicinal chemist had to do was to break the data into manageable chunks that could be progressively subdivided until a class of promising substances appeared. ClassPharmer was an automated reasoning system that did exactly that, just as a medicinal chemist would: it started with a large data set, sifted through it, and broke it down into natural classes based on underlying chemical patterns and affinities, setting aside promising classes for further analysis. It worked as their prototype system had, except that it was more reliable, accurate, powerful, and speedy.

To get its automated system to the point where it could act, think, and make judgments like a chemist would, Bassett and the other Bioreasoners had once again gone out into the field, interviewed chemists, and then refined their software in light of what they'd learned. It had paid off in the end.

"The classes that we're getting now are just absolutely superb," said Bassett. "The software automatically searches the data, finds the natural chemical families, and classifies them. That's the key, the basis of our technology, that we find chemically relevant scaffolds from which to build classes, and we find them automatically with the software."

ClassPharmer, furthermore, did not require massive servers or high-end workstations in order to do its work, but could be run on standard Windows NT computers. The user entered the screening data in the SMILES format and chose various options from among the menus with selections such as "Classes," "Search," and "Find Substructure," and the computer did the rest. A conventional desktop computer could run through two thousand compounds of molecular weights up to 800 in about an hour. The company's other automated systems, LeadPharmer and CompoundSelector, had their own separate roles to play downstream of the class-forming process, further winnowing out likely drug candidates from the initial welter of data, confusion, and noise.

By the end of 2001, Bioreason had generated more than $3.5 million in revenue through the sale of its automated reasoning systems, at prices ranging up to $100,000 for a complete package, to companies such as Warner-Lambert, Pfizer, Pharmacia, DuPont, and Amgen, among others. In addition, the company had recently opened a European division, with three employees working in an office located at 15 Avenue de l'Europe in Strasbourg, France. By any standard, this was progress.

12

THE NEW SILICON VALLEY

BY THE YEAR 2001, Daylight's software systems had achieved a truly impressive market penetration. Between 1992, when the company relocated to Santa Fe, and 2000, when it moved into its glossy new research headquarters west of the city, Daylight had grown from an out-of-the-way startup whose products were used by a small band of fervent supporters to an established firm whose software had almost completely saturated its intended market. In 1992, Daylight had approximately two hundred customers and, among them, some five hundred users of its various software systems. Ten years later, the number of companies using its software had risen to four hundred, but the number of users had swelled to about twenty thousand. Yosi Taitz, Daylight's chief executive officer, estimated that by 2001 the company had managed to get its products into 97 percent of the world's major pharmaceutical, chemical, agricultural, and biotechnology companies, which meant that Daylight's systems had become a nearly universal tool for industrial chemists, drug developers, and other researchers. For many of these people, indeed, Daylight's systems had changed the way they did chemistry.

There was, for example, the case of Bill Ellis.

Bill Ellis is a tall, gaunt man in his sixties, soft-spoken and retiring almost to the point of taciturnity. He is chief of the chemical information department of the Walter Reed Army Institute of Research in Silver Spring, Maryland, a large modern building set on the top of a hill and backed by a forest. Most of the place consists of laboratories, instruments, freezers, and animal cages.

The institute is the army's medical research branch, and since its founding it has been charged with the considerable task of protecting American soldiers against disease, particularly those conditions that drug companies considered to be "orphan" illnesses because they were practically nonexistent within American borders: malaria, leishmaniasis, pneumocystosis, and the like, all of which were of parasitic origin. Parasitic infections were so rare in the United States (particularly before the advent of AIDS) that most of the country's parasitology research was done on behalf of the animal population, not people.

Malaria was endemic in Africa, however, and in parts of the Far East, including Vietnam. For many years, chloroquine was the treatment of choice for malarial infections, but the drug had become less effective as the plasmodium that caused the illness gradually became resistant to it. During the Vietnam War, therefore, the army needed a new drug to fight chloroquine-resistant malaria and asked the Walter Reed Army Institute of Research to come up with one. The manner in which the institute's medicinal chemists responded to that request was a paradigm case of classic, pre-Daylight methods of drug discovery.

The first thing the institute's researchers did after getting the request for a new antimalarial preparation was to ask the American drug industry for an immediate and massive donation of chemicals. Basically, they requested samples of every last chemical compound that those companies had on their shelves. The compounds in question were not necessarily drugs, successful or otherwise; they were merely chemicals of whatever type, and in some cases were compounds of unknown structure, reactivity, or biological effect, if any.

"Tens of thousands of chemicals came pouring in here," Bill Ellis said. They came in various quantities and in different containers, shapes, forms, and sizes: tiny two-gram lots, small vials, brown bottles, one-liter jars. Some were liquids, some were solids, some were powders.

Still, the compounds that the drug companies had on hand represented only a small fraction of those that were chemically possible, and so in addition to their acquisition of existing chemicals, the Walter Reed Army Institute started an equally massive synthetic program, issuing contracts to chemical companies to synthesize hundreds of new compounds never before seen.

"In one year we synthesized or accessioned one way or another fifty thousand different compounds," Ellis said. At the end of this influx the Walter Reed scientists had roughly a hundred thousand compounds in storage, which they then had to have some way of keeping track of. "It's no good having a hundred thousand samples if you don't know where they are. Some have to be desiccated, some have to be refrigerated, some have to be frozen; some are highly volatile, some are poison, etcetera."

This being still the prehistoric era of record keeping, well before the advent of desktop computers and easy-to-use database programs, the scientists kept track of their samples by hand, logging them into lab notebooks—physical ledgers with pages that turned, ripped, and collected inkblots, stains, and smudges.

"That got messy in a hurry," Ellis said.

But record keeping was merely the beginning. The only way to know what those hundred thousand chemicals did, and in particular how useful they might be against the malaria parasite, was to test them all for biological activity, which is to say, by doing laboratory experiments. And so the Walter Reed scientists launched a comprehensive testing program, experimenting with the compounds first in vitro (in glassware) and then in vivo, in mice. At one point, during the peak of the antimalarial discovery program, they were putting a thousand compounds a week through wet-chemistry and small-animal trials.

For the in vivo component of the trials, each separate compound had to be evaluated for activity against a set of sixteen mice, fifteen of which were infected with the parasite and one of which wasn't, as a control. During the high point of the in vivo screening, this meant that the scientists were going through the animals at the average rate of sixteen thousand mice per week. That too produced a rather sizable data cache, one that the researchers kept, as before, in physical, handwritten lab notebooks.

Nightmare that it all was, by 1972 the Walter Reed scientists had succeeded in discovering a compound, mefloquine, that was active against *Plasmodium falciparum*, the chief malaria parasite. The researchers started human clinical trials in the same year, and after further development work done in cooperation with Hoffmann-LaRoche, mefloquine was finally approved by the FDA, at which point it became the new treatment of choice for the disease. The drug is marketed today under the trade name Lariam.

The development process had been a long slog through a barely navigable data swamp, and it was clear that the Walter Reed scientists would need more efficient techniques for inventing new drugs in the future. Specifically, they'd need a computer-based record-keeping system for inventory control, biological activity reports, and chemical structural information. In 1978 Bill Ellis set out to invent just such a beast.

He was partially successful. Inventory control was the least of his problems, for he easily developed a system of bottle numbers and associated shelf locations, using prefixes such as "R" to identify samples that were refrigerated, and so on. Tabulating the animal results, likewise, was no big problem, simply a matter of entering the relevant data alongside the bottle number.

Putting chemical structural information into the computer, by contrast, was quite another matter. Indeed, it was next to impossible. "Computers didn't know what chemistry was, they didn't know what chemicals were, you had to teach them," Ellis recalled. "You had to teach them what a carbon was, what a nitrogen was, and a hydrogen, and a whole lot of other things. You had to teach them the number of bonds for each atom, the valences, etcetera."

But by dint of huge effort and years of work, Ellis managed to cobble together a working system, a program he called "ChemStructs."

Even he was never entirely happy with his creation. For one thing the code was a chore to maintain, most of it having been written in hexadecimal, a machine-language code of ones and zeroes set on a numerical base-sixteen. Even to those who were adept at the system, hexadecimal notation was arcane in the extreme.

For another, Ellis's ChemStructs system failed to classify certain chemical structures—tautomers—correctly. Tautomers were types of molecules that shared the same formula but were structurally distinct, a situation that remained unworkable to the computer despite all of Ellis's efforts to implement it.

Worst of all, Ellis's system had no reliable way of representing the structure of individual molecules uniquely. His system mapped molecular structures onto a Cartesian coordinate system in which the location of each component atom was designated by a number that represented the atom's precise position in three-dimensional space. The problem with this, however, was that the representation of the molecule's structure then became a function of the molecule's spatial orientation. A given molecule aligned one way was, so far as the computer was concerned, a completely different animal from the same molecule rotated by any number of degrees in any other direction. His software, in short, had trouble realizing that the two different-looking structures were in fact one and the same molecule observed from two distinct points of view.

Still, the system worked after its fashion, was adopted for use by the American Chemical Society, and later became the basis for more advanced applications produced by other software developers.

In the early 1980s, however, Ellis heard of a parallel effort at Pomona College that not only had solved the problems of tautomers and that of "canonicalizing" a molecular structure irrespective of its orientation in space, but also allowed for powerful searches of many different types and even allowed the user to experiment with structure-activity relationships, changing a molecule's structure on-screen and then immediately learning how those changes affected its chemical behavior. This was Dave Weininger's SMILES nomenclature system together with its associated Thor database and Merlin search engine. Ellis was an early adopter of Daylight software and a quick convert to the company's information-based methods of doing chemistry.

"The big advantage of SMILES," Ellis said, "is that it consistently comes out with the same structure. No matter how you put the structure in, the system will create the unique structure, one that always looks the same."

Daylight's Thor database system, for its part, allowed Ellis to store virtually any arbitrary amounts of structural, biological, or other information and to call it up again in a split second. In late 2001, Ellis had information on approximately 300,000 separate chemical compounds stored in Thor, together with identifiers that fixed the location of each sample in the Walter Reed inventory housed in nearby Rockville, Maryland. And for those compounds that his scientists had tested, Ellis's Thor files stored data on the relevant in vitro and in vivo trials, along with their results. By the spring of 2002, he was hoping to take delivery of 100,000 additional samples from Russian pharmaceutical and chemical inventories. When they arrived in his lab, the available data on those compounds would be integrated with the others in the Thor database.

Useful as the Thor database was, it was the Merlin search engine that

permitted Ellis to act as a magician of computational chemistry, performing feats of chemical combination, modification, and similarity-searching
that before he could only vaguely dream about. Investigating structure-
activity relationships was the prime example of this. "If you get a compound that's active, you try to improve it, make it less toxic and more
active," he said. "This is done by playing around with the chemical structure of the molecule."

Previously, he'd had to do this in test tubes, experimenting with compounds one by one. With Merlin, Ellis does this on-screen, starting with a
base molecule—mefloquine, perhaps—and then modifying it. "You move
things around, twist this, change that, bend something, add something,
take something away. You start adding things, chlorine, methyl groups, an
acid group, whatever, hanging different pieces of 'spinach' onto a base
molecule to see how it affects the molecule, whether it makes it better or
worse.

"You sit here for a long time," he added. "You stare at a lot of structures,
and you work through the data."

If he finds something interesting, he goes back and looks in his inventory to see whether he has that compound, or a similar one, already on
hand. If not, Daylight's Reaction Toolkit will tell him how to synthesize it.
"The program will give you the reactions in question," he said. "It will tell
you the pathways to make the thing. The program will go back one step,
then a step before that, and so on."

It was while he was playing around with the structure of the mefloquine
molecule recently that Ellis found "something interesting." He had started
with the original structure of the molecule and then using GRINS (Daylight's Graphic Input of SMILES utility), had clipped off a few hydrogen
atoms from its outer edges.

"I just knocked off these little fellows here," he says, reproducing what
he'd done, "and ran a search for new molecules containing the remain-
der."

The new molecules would contain the original mefloquine structure
less the knocked-off hydrogen atoms, but they would also contain all sorts
of added molecular bits and pieces, accretions that would give the base
molecule new properties, some that made it a better drug and others that
made it worse.

"Okay, this is the structure I'm gonna run," he says, "and I go back to my control column, and I'm gonna run this against 300,000 structures that I have out there, in the database, and I'm gonna say: 'Select everything that has this structure, at least.' "

"To give you an idea of how fast it will go out and get 'em," he says, then pausing for as long as it takes to snap your fingers, "There, it's done. It searched 264,471 compounds, and of those it found 284 that are related, containing at least the original structure, plus variants."

He now starts scrolling down through the long tableau of new molecules, all of which are pictured in color.

"The original compound, the 'at-least' compound, is in yellow," he says. "The new stuff—the additions, the 'spinach'—is in blue."

Once he had gotten the list of 284 hits, he had his staff of wet-lab chemists start testing them in vitro for possible activity against bacteria. The chemists found that many of the compounds were active against Gram-positive bacteria, a vast category of microorganisms that embraced some of the worst pathogens known to humankind, including those that cause anthrax, botulism, tuberculosis, and other diseases. Suddenly it looked as if Ellis might have a valuable new antibiotic on his hands, although, as he realized, "a lot of hits are nothing more than toxins: you put them in animals and they die."

Some months afterward, in October 2001, letters containing the anthrax microbe, *Bacillus anthracis*, were turning up all over Washington. On October 25, in fact, anthrax spores turned up in the mailroom of the Walter Reed Army Institute of Research, three floors down from where Bill Ellis himself worked.

Ellis thought that just possibly, somewhere within those 284 hits was a new cure for anthrax. Shortly after the anthrax-letter attacks, Ellis arranged to send a sample of the compound to the U.S. Army Medical Research Institute of Infectious Diseases at Fort Detrick, Maryland, where researchers had the necessary biosafety equipment and lab facilities to test the drug against the disease.

All of those 284 hits were from a single Merlin search of the Army's 264,471 chemical compounds, a search that had taken, at most, half a second.

IN THE SHORT span between 1999 and 2000, Stu Kauffman originated a series of creative enterprises that even for him, a man of known propensities for starting grand new projects before finishing the old ones, was awe-inspiring to behold. First he spun off a new molecular biotechnology company, CIStem Molecular Corporation, in San Diego, California. A year later, he and some others at the BiosGroup spun off a second company, Prowess Software, located in Santa Fe. Prowess had nothing to do with biology, molecular or otherwise, but was engaged in the somewhat baroque niche specialty of applying complexity theory to mundane corporate purchasing decisions. The company's initial software release, Market-Prowess, was designed to help large corporations get the best deals on such lofty items as product containers, cardboard packaging, safety seals, advertising leaflets, and labels printed in any one or more of five languages. Kauffman, apparently, was a man who labored under no delusions as to what constituted acceptable behavior for a high-level scientific genius, particularly if substantial cash returns were in the offing.

Still, as if laying to rest any suspicion that he had somehow lost his scientific cool over such corporate trivia as the purchasing of shrink-wrap, glue, and soy ink, Stu Kauffman, simultaneously with the founding of these two new offshoots, undertook to write and publish a new book, *Investigations*, containing the very latest of his metascientific and semiphilosophical reflections. Intentionally titled as a tribute to the philosopher Ludwig Wittgenstein's *Philosophical Investigations*, itself a landmark expanse of gnomic wisdom, Kauffman's book was a personal milestone for him in that even he, the author, was to a large extent baffled by its contents. "Having completed *Investigations*," he said in the preface, "I remain profoundly puzzled by what I have said."

And with reason. For, as he added, "Whatever *Investigations* is—useful, as I hope, or foolish—it is not normal science."

What, then, was it? Among other things, it was a proposal for a new way of thinking about science, life, and the interrelationship between the universe and the biological organisms within it. Organisms and the cosmos, Kauffman said, were "coconstructing" phenomena: each of them influenced, controlled, and prepared the way for the further emergence and development of the other.

Above all, the book questioned the very assumption on which normal science was built, the presupposition that science proceeded by prestating the conditions of a given problem, performing numerical calculations based on known laws, and discovering the correct answer as a consequence. The difficulty with that approach, Kauffman said, was that this classical, orthodox, and ostensibly all-embracing procedure just wouldn't work when applied to biology. It was not so much that biologists couldn't prestate the initial conditions of a situation; the problem was that even if they could, it wouldn't help, the reason being that with living organisms the starting points did not lead in any fixed, deterministic, and predictable way to a specific outcome. Instead, owing to the role played by chance, random processes, and natural selection, initial conditions regularly gave rise to types of evolved organisms whose conformations nobody could have foreseen at the beginning.

But if you couldn't predict the evolution of even a single organism, you could do no better with the biosphere as a whole. "The biosphere, it seems, in its persistent evolution, is doing something literally incalculable, nonalgorithmic, and outside our capacity to predict, not due to quantum uncertainty alone, not deterministic chaos alone, but for a different, equally, or more profound reason: Emergent and persistent creativity in the universe is real."

The worst of it was that what was true of the biosphere was also true of the universe as a whole, Kauffman said, for "if we cannot prestate the configuration space of a biosphere, how can we prestate the configuration space of the universe?" The whole was no longer reducible in any simple way to its parts, and suddenly the ability of classical science to know the universe it claimed to describe was being called into question.

No, this was not "normal science," and Stu Kauffman himself didn't know exactly what to make of it all, for as both a physician and a biophysicist, he had been trained in the tenets of reductionism, a view of science that he was not willing to abandon.

"I think I've come up with a new, funny limitation to the way we've been doing science," he said. "And I'm deeply unsettled by this. I'm a Western-trained scientist, and I'm silenced by my own goddamn book. It rubs me the wrong way."

Still, this new book of his was off in its own separate intellectual compartment, for to Stu Kauffman, a man of competing theoretical, practical, and entrepreneurial interests, none of these epistemological-metaphysical problems interfered in the least with his ability to conduct business.

By the time *Investigations* came out, in fact, it looked like Stu Kauffman was at least two persons in one, for he was a man spilling over with ideas, sciences, technologies, simulations, companies, projects, and patents. One of those patents, issued in September 2000, was for a "system and method for the synthesis of an economic web and the identification of new market niches" (U.S. patent no. 6,125,351). The patent made reference to Stu's biological text *The Origins of Order* (as well as to Wittgenstein's *Philosophical Investigations*), thus fulfilling Larry Wood's vision of the book as being "really about business and management."

There was far too much creativity here, too many separate flights of the imagination, too much advancement along the frontiers of too many scientific disciplines, too many companies, too much software to be the output of one person. What single individual could have played a starring role in the launching of complexity science, could have founded the different companies, hired the staff, run the labs, taught the students, directed the graduate projects, authored or coauthored the hundred-odd technical papers, done the simulations, filed for the dozen or so patents, and written the three books? And that list does not even include his various leisure pursuits, for on top of all this Stu Kauffman skied, he hiked, he climbed mountains (he'd gotten some of his best thoughts for *Investigations*, he said, "after a month trekking in the Everest region"), he sailed boats, he had kids, and he bought and sold Santa Fe real estate. Indeed, the scale and scope of all this fevered activity was fairly exhausting even to contemplate.

Viewing this excess movement at close range, a short list of Info Mesans began to wonder whether Kauffman was not in fact spreading himself too thin, whether there wasn't an accompanying loss of depth to his overall intellectual program.

"I just don't think they're very grounded scientifically," said one Info Mesan about the BiosGroup.

"Stu thinks he's a scientist, I think he's P. T. Barnum," said a Santa Fe scientist who once shared a speaking platform with Kauffman. "It hasn't come to a shouting match yet, but at a recent lunch, held for the Spanish minister of science, he stood up and gave his two-minute talk about Bios for ten minutes, wherein he waffled on about 'moving beyond the reductionism of twentieth-century science to a new understanding of organisms as a whole,' which of course made my bullshit meter hit eleven."

"Stu is the perfect Santa Fe scientist," said another Info Mesan. In light of the city's New Age reputation, this was somewhat of a double-edged comment.

But Stu was prepared for faultfinding, and all such carpings seemed to bounce harmlessly off his hard hide. Once during a formal debate with him, the evolutionary biologist John Maynard Smith cracked some jokes about Kauffman's endless computer modeling as being a form of "fact-free science." Instead, it was all theory.

"My problem with Santa Fe," said Maynard Smith, "is that I can spend a whole week there and not hear a single fact."

"Now that's a fact!" Kauffman replied, unfazed by it all.

He lived in a fabulous house on Camino Cruz Blanca, a narrow dirt road high above Santa Fe. He enjoyed good food, expensive restaurants, and fine wines. He drove an ice-blue Jaguar convertible to work and had the best parking space at the BiosGroup. His books sold. His company had blossomed and prospered. He invented things. He made money. He had fun. What more could he want?

Only more of the same, apparently. Stu's first Bios spin-off, CIStem Molecular, harked back to Kauffman's early interest in the field of gene expression and genetic regulatory networks, work that had led to his first major use of computer simulations. The purpose of the new company was to utilize the phenomenon of gene expression as a basis for developing novel treatments for a variety of diseases.

Many diseases were at least partially caused by one or more defective genes, or defects in the expression of an intact gene. Whereas most medical interventions worked downstream of the expressed gene, CIStem's goal was to intervene at the level of the gene itself, for the treatments it pioneered would prevent the gene from expressing itself in such a way that caused disease.

The name "CIStem" referred to *cis* sites, regions of DNA that controlled the transcription and expression of nearby gene sequences. The idea was to identify certain *cis* sites of interest and then introduce molecular mechanisms that would modify the nearby gene's activity, turning a harmful gene sequence into a benign one. With such an ability you could alter the genes that controlled cell generation, tissue differentiation, and so forth, leading to new treatments for cancer, diabetes, and neurological diseases, plus advances in bone repair and

cartilage replacement, among other things. But first you needed a way
of finding the right *cis* sites.

One time when he was in Switzerland hiking in the mountains with
Marc Ballivet, the latter came up with an idea for identifying *cis* sites
essentially on a production-line basis. So Ballivet and Kauffman once
again filed for a patent on the new technology. The patent, "method of
identifying cis acting nucleic acid elements" (U.S. patent no. 6,100,035)
was awarded in August 2000, just a month before Stu received his patent
on a method for discovering new market niches.

Kauffman founded CIStem Molecular to commercialize the technology
and then set out on a nationwide search for a company director. In 1999
he hired Anne Crossway, who had a Ph.D. and M.B.A. and was the former
CEO of Cosmederm Technologies. Since Crossway was living in San
Diego, which was already a prime biotechnology hub, Kauffman set up
the company there instead of in Santa Fe. Soon afterward, Crossway
coined the term *regulomics* to refer to what the company was engaged in
doing: regulating cellular mechanisms at the genetic level.

Far-out and unproved as all of this was, CIStem had no shortage of
potential customers. One of the first was a seed company, Pioneer Hi-
Bred International, a division of DuPont, representatives of which visited
the Bios offices early in 2001. The company owned certain corn-seed lines
that produced low outputs of corn oil per kernel and other lines that gave
high corn-oil outputs, and they wanted to understand the gene-regulating
factors responsible for producing the two opposed yields.

"They are perfect for CIStem," Kauffman said. "CIStem can walk into
the low and the high lines, pull off our trick, see if we can pull out the *cis*
sites in the low versus the high lines."

Later, CIStem undertook a collaboration with researchers at Thomas
Jefferson University in Philadelphia and others at the University of
Delaware, and in 2001 the group was awarded a three-year grant from the
Defense Advanced Research Projects Agency of the U.S. government to
profile the genetic network underlying a neurological system.

The BiosGroup's other offshoot, Prowess Software, was slightly less
ambitious in its goals: all it wanted to do was to optimize a given company's
purchasing decisions. There was nothing especially glamorous about buying supplies and raw materials, but when these were done on the vast
scales of some of the world's biggest corporations, then there was a con-

siderable amount at stake, and more than just money. If you were Ford Motor Company, for example, and you needed to buy parts for a million brake assemblies, then all kinds of factors entered the picture—supplier reliability, quality control, defect rate, lead times, delivery dates, service levels, shipping costs, tiered pricing, and for international suppliers, tariffs and currency exchange rates—and all these various "dimensions" added up to a sum total. It took lots of time, energy, and brain power for a purchaser to assess the interplay of all those complex factors manually, and the MarketProwess software was conceived as a means of automating that process to the maximum extent possible.

Its creators had in fact conceived of the software as a giant brain (and had named a submodule of it "MarketBrain"), for the system would go so far as to interrogate the user about such subjective factors as personal stylistic preferences in dealing with suppliers. Then, with all of the purchaser's requirements and preferences in place, the system would execute various "procurement scenarios" over the Internet. All of this prowess would not be cheap, and the price of customized MarketProwess installation would be somewhere between $500,000 and $1 million.

In July 2000, Bios established a partnership with MetalMaker, a Chicago-based Web hub for foundries, steel mills, and their suppliers, who would give the new software a trial run. In September, the Ford Motor Company, which was also to use the new software, acquired an equity stake in the BiosGroup. A month later, Prowess Software was operating out of its own office space on Harkle Street.

A FEW DAYS into the new millennium, Stu Kauffman told a journalist from *Wired* magazine, "I'm told that Santa Fe feels like what Silicon Valley felt like ten years ago."

Kauffman was not alone in that impression. The term *Info Mesa* had been coined in the late 1990s as a deliberate backhanded reference to Silicon Valley.

"I was looking for a name for us, meaning *Not Silicon Valley*," Dave Weininger recalled. "Janet Newman, a crystallographer from Australia who was then a postdoc at Los Alamos, was over for tea one afternoon, and among her suggestions was *Info Mesa*."

To Dave the name seemed exactly right, as it vividly captured the essence and spirit of the automated knowledge-discovery revolution that was taking place on the Santa Fe plateau. He volunteered to create and maintain an Info Mesa home page (daylight.com/infomesa), and at the beginning of the year 2000 the page listed fourteen companies. In alphabetical order they were: Bioreason, the BiosGroup, Complexica, Daylight, Genzyme Genetics, Metaphorics, Molecular Informatics, the National Center for Genome Resources, OpenEye Scientific Software, PHASE-1 Molecular Toxicology, the Prediction Company, the Santa Fe Institute, Strategic Analytics, and the Swarm Corporation, all of them in Santa Fe.

Only two years later, Daylight's Info Mesa Web page listed twenty-five companies, including all of the original fourteen plus eleven new ones: Adaptive Network Solutions Research, Bionic Arts, CommodiCast, DNA Mining Informatics Software, eOrder$ource, Flow Science, Mesa Analytics and Computing, Situ Partners, Symtezzi, Turbo Linux/Turbo Labs, and Veridian Systems. Even so, Dave's new list was not complete: it left out Assuratech ("Scientific Solutions for Managing Capital Risk Exposures") and QTL Biosystems, which was developing molecular biosensors. Molecular Informatics, meanwhile, had merged with Applied Biosystems.

In 2002, moreover, the Info Mesa's firms spilled out beyond the city limits of Santa Fe. DNA Mining Informatics was in Tesuque, some ten miles north, while Adaptive Network Solutions Research was located in Los Alamos. There was a measure of poetic justice in this, Los Alamos being what it was to the history of the science and technology of the area.

Collectively, these twenty-six companies, most of which did not even exist a decade earlier, were now doing perhaps a hundred million dollars worth of business annually. This was nothing like Silicon Valley's billions, but the Info Mesa was still young, and the amount of commerce was enough to foster the perception that Santa Fe was about to become a new technological boomtown, a sort of Silicon Valley Southwest. In the year 2000, indeed, there were about the same number of new startups on the Info Mesa as there had been in the Silicon Valley of the late 1960s.

In fact there were several parallels between the two regions. Both Silicon Valley and the Info Mesa were places where small fortunes could be and were made in a short time. Both of them had invented the newest-wave creations in technology and applied science, and made money by exploiting them. Both were centers of extremely high brain wattage, innovation, risk taking, and entrepreneurial zeal, and both were flush with rank individualists, some of whom were furnished with outsized egos.

Above all, there was a realization in both places that something new, different, and fundamentally important was going on within their borders, and that their numerous enterprises giving birth to hot new technologies were in the process of ushering in some sort of as yet undefined radical transformation in the world at large. In its heyday, Silicon Valley had been legendary for corporate visionaries—Steve Jobs, for example—talking about "changing the world," "making a dent in the universe," and the like, messianic rhetoric of a sort that certain select Info Mesans—George Cowan, Murray Gell-Mann, Stu Kauffman, Dave Weininger, Anthony Nicholls—were not above engaging in themselves. And so with all these common elements uniting them, it really did seem as if Santa Fe were about to emerge as America's new Silicon Valley.

The Info Mesa, however, was Silicon Valley with a difference.

Silicon Valley had begun as such in 1955 when William Shockley, coinventor of the transistor, founded Shockley Semiconductor Laboratory in a concrete-block warehouse on South San Antonio Road in Mountain View, California. Two of the lab's employees were Robert Noyce and Gordon Moore.

The mission of Shockley's lab was to investigate the properties of semiconductors, solid crystalline materials such as germanium and silicon whose ability to conduct electricity could be modified by the addition of minute quantities of specific chemicals (aluminum, phosphorus, boron, and arsenic), a process called *doping*. These doped materials were of interest to scientists because when small flakes of them were attached to electrodes and then stimulated with pulses of electricity, they could be made to act as electrical switches, allowing electrical currents either to pass through them or to stop dead, depending on the polarity of the current, its voltage, and other factors. The crucial aspect of this phenomenon was that each tiny square of silicon was acting as a switch, even though it wasn't a switch in the conventional sense, which was to say that it wasn't a mechanism with moving parts that opened and closed a circuit. Instead, a semiconductor permitted or prevented the flow of electricity by means of alterations in the very structure of the atoms, molecules, and free electrons of which it was composed.

Because its component parts were small batches of molecules rather

than macroscale devices, semiconductors could switch between on and off
states at absolutely unbelievable speeds, rates that mechanical devices
could never remotely hope to approach, on the order of millions of cycles
per second. Semiconducting materials, therefore, constituted the ideal
medium for the creation of small and fast desktop computers. Thin, intersecting lines of silicon could be arrayed in complex patterns on integrated
circuits ("computer chips"), the chips could be arranged together on a circuit board to perform specific functions, and stacks of those boards,
together with monitors, input/output devices, and so on, could be fash-
ioned into what would become the basic personal computer.

The Shockley Semiconductor Laboratory lab never produced any of
these things—Shockley, apparently, was a difficult man to work for and
was his own worst enemy—but his employees did, once they left the place
and started companies of their own. In 1957 Robert Noyce and Gordon
Moore got financial backing from Fairchild Camera and Instrument, and
founded Fairchild Semiconductor. Fairchild Semiconductor was Silicon
Valley's first successful semiconductor firm and in turn spawned several
others of its type. Noyce, simultaneously with Jack Kilby of Texas Instruments, invented the integrated circuit, and later Noyce and Moore
founded the Intel Corporation to manufacture a densely packed and compact version, the microprocessor, in vast quantities. Intel, together with
more than fifty other Fairchild spin-offs (including National Semiconductor, Advanced Micro Devices, Signetics, General Microelectronics, Inter-
sil, Qualidyne, and Siliconix, among others), soon transformed what had
once been 100,000 acres of prime agricultural land south of San Francisco
into "Silicon Valley," thereby causing the fabled upward spiral of growth in
factories, housing, and real estate prices. At the end of it, Silicon Valley's
squarish prefabricated buildings and street patterns grew to resemble the
very integrated circuits that their firms poured out by the hundreds of
thousands.

But here lay the critical difference between Silicon Valley and the Info
Mesa. Silicon Valley existed on the creation and sale of physical objects,
specifically, *electronics*: computer chips, circuit boards, modems, disk
drives, monitors, printers, plotters, scanners, and other peripherals. The
only element of a digital computer that was not itself a physical object was
the software that ran inside it. But except for Apple Computer, which produced its own operating systems, Silicon Valley mostly did not deal in soft-

ware, much of which came from Microsoft, which was based in Washington State.

The Info Mesa's firms, by contrast, manufactured no physical object of any description; instead, they traded exclusively in data, software, information, and knowledge. This meant that Santa Fe had no need for factories, assembly plants, production lines, equipment, tools, or vast hordes of workers. Nor did its collection of applied-science businesses bring to the area any of the congestion, housing problems, or pollution that normally accompanied the large-scale expansion of industry. The Info Mesa was a Silicon Valley without the sprawl, ugliness, or environmental impact of the hitherto conventional boomtown.

The Info Mesa's principal output, scientific knowledge, had no inherent size, shape, scale, or bulk. Whether in the form of software systems or bodies of finished information, the whole of a company's output could be sent out on a compact disc or transmitted invisibly over the Internet. This was a form of spotless commerce, based on the creation, sale, and transfer of wares that were impalpable and intangible and possessed neither height, weight, form, nor dimension.

The business of the Info Mesa, therefore, was transacted by companies housed in either single buildings or suites of offices. The Prediction Company, Bios, Daylight, the National Center for Genome Resources, and Swarm all operated out of single, freestanding structures. Bioreason, OpenEye, and many other companies merely rented office space.

The gold rush taking place on the Info Mesa was of an extremely polite and civilized variety, without factories, warehouses, and vast parking lots, which meant that despite the variety and number of science-based businesses that sprang up in and around it, Santa Fe could remain essentially the same place it had always been: laid-back and somewhat sleepy, a charming tourist town, albeit stylish, expensive, and awash in new capital.

It was Silicon Valley, the stealth version.

13

THE ECLIPSE OF NATURE

KNOWLEDGE DISCOVERY BEING the main business of the Info Mesa, it followed that when vast new data caches arose that could be turned into knowledge, Santa Fe's informatics firms would be there to effect the transformation. Such was the case with genomics, the attempt to make sense of and to use for human benefit the genetic data contained in human, animal, and plant DNA.

The Info Mesa entered the genomics business in 1994, when the National Center for Genome Resources (NCGR) moved into offices on Old Pecos Trail, just south of the city proper. The center was to be a clearinghouse for genomic information but was to be a nonprofit institution, one that would make its resources freely available to the public. Originally, the NCGR had owned a for-profit subdivision, Molecular Informatics, a firm that specialized in providing chemical information services to drug, chemical, petrochemical, and agricultural firms. In 1997, the mother company sold off Molecular Informatics and used the proceeds as a general endowment fund. NCGR got additional contributions from private industry, the federal government, and various foundations, and in the spring of 1999 announced that it would build a thirty-two-thousand-square-foot office complex in the new Rodeo Business Park on the southern outskirts of Santa Fe.

When it was formally dedicated in April 2000, NCGR had become the Info Mesa's chief proponent of using human, animal, and plant genomics for improving human nutrition and health and cleaning up the environment. At that time the center employed some sixty scientists, including geneticists, computational biologists, and biochemists, among other specialists. It also owned and operated a sixty-four-processor Sun Enterprise 10000 high-performance computer, a system that ranked in the top two hundred of the world's most powerful supercomputers in use at the time. The center's researchers worked on a variety of projects ranging from helping sheepherders in nearby Española select the best species of sheep to use for certain breeding purposes, to studying the genomes of different types of grasses with a view to strengthening them, to the construction of a comparative genome-mapping tool, a piece of software that would allow scientists to learn about one species of organism by using genomic knowledge that pertained to another.

The grass project, financed by a $2 million grant from the National Science Foundation, held out the prospect of increasing and improving some of the world's basic foods through the comparative study of plant DNA sequences. The "grasses" in question were barley, maize, rice, sorghum, and wheat, five closely related crops that in one form or another constituted a large portion of the world's food supply. The goal of the project was to increase the hardiness, yield, and nutritive value of these plants "bioinformatically," not by growing them experimentally on the farm but rather by reading, understanding, and perhaps modifying their respective genomes—the specific gene sequences that made each plant the distinct biological entity it was.

The scientists approached the problem by generating stretches of the DNA sequences of each plant and then analyzing them for content, organization, and the presence or absence of other types of DNA. One of the tools they used in the work was NCGR's "Genome Feature Viewer," the center's genome browser. With it, the user could view DNA sequences as if they were Web pages, complete with graphs, charts, and other images that could be manipulated by means of check-boxes or pull-down menus or merely by clicking on various parts of a genome map. After spending a certain amount of time browsing the relevant sequences, investigators would be on as intimate terms with small, selected regions of a given genome as they were with their own offices. Then by comparative DNA analysis of the same region in other plants, a researcher could identify the genes that conferred a certain desired genetic trait—drought resistance, for example.

At that point, the researcher could entertain the thought of incorporating into the second strain the genetic sequences that conferred drought resistance to the first strain. If that were done, then just possibly the second variety of plant would be as drought resistant as the first.

The general objective, at any rate, would be to use the knowledge gained in these studies to increase crop yields and to make the crops more resistant to bacterial pathogens, viral diseases, or insects, or able to thrive in colder or hotter, drier or wetter environments, and the like. Such was the power and leverage of genomics—of reducing organisms to biological information—that by spending a matter of days or weeks at the computer, a researcher could discover a method of improving a given crop that might take farmers years to achieve by conventional cross-breeding and trial-and-error methods, if indeed the result could be obtained by those means at all.

The NCGR's research program was not confined to plants, however. The center also maintained a variety of genomics databases on subjects as diverse as plant pests and metabolic reactions across several different biological species, and gene expression libraries on bacteria, plants, and humans. Much of this information was freely available on NCGR's Web site and open to any qualified user.

"Our vision is to get as much scientific data into the scientific public domain as possible, which is why our information is on the Web," said Stephen Joseph, NCGR's president. "This approach of working with data is only four or five years old. Agriculture, medicine, and biology in general will be revolutionized. Things will be very different in the future."

PHASE-1 Molecular Toxicology, another Info Mesa firm, was changing the way toxicology was done by transforming it into an information science as opposed to an experimental one. Instead of injecting animals with a substance to determine whether it was poisonous or harmless, the company would utilize more knowledge-based and humane methods. The traditional animal-injection method didn't always work very well in practice, and every so often a drug made it to the marketplace with worse side effects than expected, as was illustrated by the weight-loss drug Fen-Phen and the heartburn medicine Propulsid, which its maker, Johnson & Johnson, belatedly withdrew from the market, causing the company stock to drop by 10 percent.

PHASE-1 Molecular Toxicology would establish toxicity by the adroit use of chemical data. A drug researcher who had identified a promising drug compound could send in a sample of the substance, and PHASE-1 technicians would assay it for toxicity by placing the compound on a chemically reactive microarray (a microprocessor programmed to identify chemicals) and then merely reading off the results. Alternatively, the researcher could provide PHASE-1 with the compound's chemical structure, and then by reference to their large databases of toxic substances, the PHASE-1 scientists could determine which of the substances were likely to be toxic to humans or animals. Further into the future, the company hoped to refine the technology to the point that a simple doctor's-office test could show whether a given drug was going to be toxic or helpful to an individual patient.

By the end of 2001, PHASE-1 had contracts worth some $9 million with twenty-five drug companies. The company employed seventeen technicians, had a full line of computers and analytical tools in their seventeen-thousand-square-foot facility located in Santa Fe's Rodeo Business Park, and had opened a subsidiary branch in Belgium.

Other Info Mesa firms exploited various other scientific niche sectors. Flow Science, a Santa Fe company with roots at Los Alamos, specialized in fluid flow. Fluid-flow analysis was an old and venerable science, textbooks having been written on the subject by Bernoulli in 1738 and by d'Alembert in 1744. But a mathematical analysis of hydrodynamics was one thing; active simulations of fluids moving through space were another. In 1963, while at the Los Alamos Scientific Laboratory, Dr. C. W. Hirt invented an approach to fluid dynamics, the so-called volume of fluid method, that would prove to be extremely adaptable to computer modeling. Later, in 1980, Hirt founded Flow Science to develop a high-fidelity fluid dynamics program, and in 1985 he released his star application, a piece of software he called FLOW-3D.

Some fifteen years later, FLOW-3D was in Release 7.7, and the company, with customers all over the world and with sales representatives in about a dozen countries, had moved into new offices on Harkle Road on the south side of Santa Fe. By this time Hirt and his associates had optimized his software to the point that it could model virtually any scientific, engineering, or other problem in fluid dynamics, including such varied phenomena as the spread of acoustic waves through various media, the solidification or shrinkage of metals, liquid flow through porous substances, the forces involved in surface tension, the sticking properties of adhesives, the laminar or turbulent flow of air over an airplane wing, the passage of ink from an ink-jet printer head, waste sedimentation rates in water-treatment plants, and snow-drift patterns around a group of apartment buildings. Anyone with a fluid-flow problem not yet addressed by FLOW-3D Version 7.7 was invited to contact the company, whose officers claimed that "it is, very often, possible to modify the program to make the seemingly impossible possible."

A company called Assuratech was applying complexity theory–based simulations to the problem of capital risk exposure. The firm's scientists had created "intelligent software systems" that, they said, allowed their clients "to test business strategies *in silico* before committing resources."

Another, Complexica, had developed a simulation for the use of insurance businesses. The program, called Insurance World, supposedly simulated the entire global insurance industry. Roger Jones, the company's chief scientist, said, "The Insurance World software captures what all the entities are doing and serves as a kind of brain prosthesis for insurers, expanding their intuition so they can adapt to surprises and survive in a complicated environment."

A firm called eOrder$ource specialized in automated ordering, inventory tracking, and record keeping for medium-size commercial firms. The company's first product, a piece of Palm Pilot–compatible software called HiPhive Rep, substantially removed human beings from the order-entry, sales-data, and commission-reconciliation loop. "One client," the company claimed, "was able to load 21,000 items from 20 different manufacturers and 9,500 customers, sorted by 'Bill To' and 'Ship To' addresses for seven sales representatives."

And then, finally, there were the one-, two-, or three-person startups, mom-and-pop scientific software outfits operating out of a garage, home office, or basement whose employees, in the finest Silicon Valley tradition, hoped to craft one brilliant notion into a hugely profitable commodity. There was DNA Mining Informatics Software, in Tesuque, whose chief product, Setter, did exceptionally fast human/mouse nucleotide-sequence comparisons. "Anywhere on the drug pipeline, people want to compare human to mouse DNA," said Bill Bruno, biophysicist, inventor of the program and the company's sole employee. His software would do it for you, quickly and automatically.

There was Los Alamos Computers, whose two employees, Gary Sandine and John E. Pearson, both of them Los Alamos National Laboratory veterans, were selling a line of made-to-order Linux-based computers to sophisticated users in academia, industry, and the government. (Linux was an operating system developed in 1991 by Linus Torvalds.) Its products were custom-tailored to satisfy unusual requirements. In 2001, for example, a chemist needed a specialized Linux system in order to run an exotic computer code he'd written. Sandine and Pearson learned the code, built a custom system around it, and sold it to the chemist for $20,000.

And there was Adaptive Network Solutions Research, also of Los Alamos, which billed itself as an all-purpose problem-solving, question-answering firm, a sort of private-detective agency for scientists, a company that would take on all-comers and all problems. Despite its small size, the firm had an impressive client roster, and the company's president and chief and sole scientist, one William C. Mead, another Los Alamos lab escapee, had done problem-solving work for customers that included the U.S. Department of Education, the Lawrence Livermore National Laboratory, and Impulse Devices, a private company engaged in nuclear-fusion research and development.

THE INFO MESANS were not the only ones in the business of automated knowledge discovery, however, and by the turn of the millennium, the enterprise of turning data caches into knowledge had blossomed to the point where it warranted and had received its own new jargon. The mother discipline was "informatics," the technology of reducing reality to information. The daughter disciplines were bioinformatics, proteomics, chemoinformatics, genomics, and regulomics, each of which would ultimately have a global range and scope.

Bioinformatics was the marriage of computer science and biology, the very fusion that Harold Morowitz and Temple Smith were trying to effect in their Matrix of Biological Knowledge Workshop back in 1987. By 2000 the two enterprises had been successfully joined forevermore, and bioinformatics courses were being taught at the Rockefeller University in New York and at the UCLA Bioinformatics Institute in California. In addition, there were government-sponsored bioinformatics centers, institutes, data banks, and other resources in the United States, Canada, England, Switzerland, and Japan. Suddenly it seemed as if all the world's biologists were viewing their pet organisms not through microscopes but through the type of computer analyses and simulations that Morowitz, Smith, Anthony Nicholls, Teresa Strzelecka, and the others had been groping toward in Santa Fe.

Genomics had become an industry unto itself by the turn of the century. Celera Genomics, in Rockville, Maryland, was a focal point of the effort to decode and decipher the human genome, the full sequence of DNA base-pairs that made up the human genetic code. This had been done by June 2000, when Craig Venter of Celera and Francis Collins, the head of the Human Genome Project of the National Institutes of Health, announced that they had substantially sequenced the entire human genome.

Still, the sequenced genome represented nothing more than blots on a page, or a series of endless *A*'s, *C*'s, *T*'s, and *G*'s, depending on the method used to display the data, and the final result was a database of the typical colossal size. At the end of its sequencing effort, Celera had 50 terabytes of data in storage, equivalent to eighty thousand compact discs, all of which, if stacked one to the next in their plastic jewel boxes, would take up almost a half a mile of shelf space. Thus the far more challenging project still lay ahead, which was to understand what the data actually meant, figuring out which exact sequences of molecules were responsible for which functions, structures, or reactions in the body.

"Methods have evolved to the point that you can generate lots of information," said Michael R. Fannon, of Human Genome Sciences, another genomics firm in Rockville. "But we don't know how important that information is."

"It's a tsunami of information," said Celera Vice President Gene Myers. "For the next two to three years, the amount of information will be phenomenal, and everyone will be overwhelmed by it. The race and competition will be who can mine it best. There will be such a wealth of riches."

It was proteomics, however, that represented the latest informatics trend.

"The terms 'proteome' and 'proteomics' came into common parlance between 1995 and 1998 by analogy with 'genome' and 'genomics,' " said the Nobel laureate Joshua Lederberg in 2001. "They do not yet appear in standard dictionaries. There's no doubt, however, that proteomics is going to be one of the next big newsmakers in biotechnology."

The object of proteomics was to identify and characterize the human proteome, the total collection of all the proteins expressed by all the genes of the human genome. The ultimate purpose of that in turn was to allow scientists to zero in from that vast collection and locate the discrete subset of proteins that were involved in healthy tissues and in human disease processes. Once a protein's role in a disease was known, then scientists could tailor drugs or other treatments to cure or alleviate the illness.

Because of the fact that proteins could be patented, the companies holding patents on proteins that played a role in diseases would be in possession of a highly valuable property. In the year 2000, Oxford Glyco-Sciences, of the United Kingdom, attempted to corner the market on certain proteins, and after raising $50 million in an offering on the London stock exchange, it filed patent applications on more than eight hundred different protein molecules. Since there were only so many proteins in the human proteome, this was analogous to a major biological land grab, and competing companies quickly followed suit, including Glaxo-Wellcome in the United States and Cellzome in Heidelberg, Germany.

In chemoinformatics, the field that Dave Weininger had done so much to advance, Daylight Chemical Information Systems by no means had the field to itself, and by 2000, Daylight had more than twenty-five competitors supplying software for molecular modeling, the creation of chemical databases, and supporting research in structural biochemistry. In the United States these included MDL (Molecular Design Limited) in California and Tripos in St. Louis, Missouri, and in Canada, the Chemical Computing Group.

What all this meant in practical terms was that by the dawn of the twenty-first century, large portions of empirical reality had been reduced to data, or soon would be. Atoms, molecules, chemicals, proteins, genes, cells, organisms, entire ecological systems, the planet itself, the solar system, the Milky Way Galaxy—all of them had been abstracted away from their particular manifestations and reduced to data, the data turned into information, and the information turned into abstract theory and conceptual knowledge. It was almost as if reality itself could be skirted, bypassed, or methodologically suppressed, so much of it had been rendered as numbers, data, and computer simulations.

There were now simulations of everything, whatever the subject, at whatever imaginable level of scale: simulations of human population dynamics, industries, even the entire global economy. There were computer simulations of thunderstorms, the progress of hurricanes, and climatological change. There were simulations of river flows, the tides, star formation, the movement of planets and their satellites, the motions of galaxies, the birth, evolution, and ultimate heat death of the universe. And except for the still hopeless problem of everyday weather prediction, most of these simulations actually worked—they made detailed predictions that were verified by observation.

The human genome had been read, chromosomes had been mapped and filed away into databases, genes had been translated into their sequences and deposited into gene banks, and the cell's regulatory mechanisms had been converted into abstract numerical functions and reconstructed in computers. Proteins had been reduced to amino acid sequences, and the molecules and their electrostatic surfaces had been visualized on display screens.

Drugs had been reduced to their chemical constituents, and they in turn had been expressed as SMILES, GRINS, or other computer representations, and the work of traditional wet-chemistry labs had been replaced by accurate simulations of chemical reactions. In Tesuque, north of Santa Fe, Dick Cramer, a researcher working in his home office, had created a virtual library of 26 trillion chemical compounds built on only seven different chemical reactions, and had stored them on the hard disk of his personal computer, a Sun workstation that could search through all 26 trillion of the virtual compounds overnight.

By the turn of the century, biology had become bioinformatics; genes, genomics; proteins, proteomics; chemistry, chemoinformatics; and cellular regulation, regulomics. The sole significant holdout to the informatics trend seemed to be elementary particle physics, where, although there were plenty of models, theories, and simulations available, a large piece of apparatus still reigned supreme: the linear accelerator, such as was still in use at the Stanford Linear Accelerator Center (SLAC), Fermilab, or CERN.

Otherwise, everything was information. External reality had become a mere touchstone, yet another data point.

Reality was still there, of course, there was no doubt about that, and it represented the one inescapable and ineradicable element of the knowledge-discovery process. It was where the data came from, and it was what any idea, theory, model, or simulation had to correspond to in order to be successful, real, accurate, and true. Reality was the sole preexisting independent variable, the only thing that, in and of itself, could confirm or invalidate a theoretical construct of whatever type.

But reality was no longer irreducible: it *had* been reduced, time and again, to the play of numbers, concepts, theories, models, and simulations. And other than its role as the original source of data and the ultimate arbiter of the truth of any human attempt to capture it by means of abstractions, reality was largely dispensable and out of sight, and you had the feeling that sooner or later it would be banished from the scene—or at least from the everyday ken of many working scientists. Methodologically, information would have eclipsed it.

EPILOGUE

THE BUSINESS DOWNTURN that swept in with the new century had not perceptibly affected the Info Mesa when Daylight held its fifteenth annual MUG, the meeting of the MedChem User's Group, that the company had been hosting since 1986. The 2001 installment was highly symbolic to Dave Weininger, who built the MUG's computer graphics around images from the Stanley Kubrick movie *2001: A Space Odyssey*, and so everything about the meeting was forward-looking and upbeat. Daylight had released its latest software in Oracle cartridge format, which was a major new convenience for users. The company's newest chemoinformatics package featuring traditional Chinese medicines, the so-called "TCM Relational Knowledgebase," was up and running in demo version. And as always, there were several other new chemical information marvels in the offing.

And so between March 6 and 9, 2001, about a hundred chemists and drug developers from around the country met at the Eldorado Hotel in Santa Fe for three days of lectures, new product demonstrations, and talks. Jeremy Yang, Daylight's chief engineer and the lead speaker, announced that Daylight's official corporate goal was to amass, tabulate, and provide "all the chemical information in the world," an ambition that, grandiose as it was, no one in the audience snickered at in the least. John Bradshaw, of Daylight's European branch in Cambridge, England, gave a "brief history of chemical nomenclature," an illustrated account in which SMILES and its miscellaneous offshoots, including SMIRKS, SMARTS, GRINS, CHUCKLES, and CHORTLES all made their appearances.

But it was Dave's own talk about the new traditional Chinese medicine database that moved into what was, for Western chemists, decidedly new territory. He'd first seen a textbook on Chinese medicines about a year previously, he said, "and I was amazed that I recognized about 80 percent of the structures pictured. I also realized that this was a storehouse of three thousand years worth of clinical trials on more than a billion subjects." It was, therefore, a subject that Daylight had to pursue. The TCM Relational Knowledgebase was the result.

Put together with the help of computational chemists at the Chinese Academy of Sciences in Beijing, the database contained chemical information, structures, names, formulas, atomic weights, and other data on 6,800 compounds, 1,268 medicines, and 1,548 plant species that had been used by Chinese healers over the ages. The project, Dave said, had been something of a nightmare since the original data had been in so many different languages, with plant species names in Latin, names of medicines and disease descriptions in Mandarin Chinese or Pinyin, and so on. All these various terms and descriptions had to be converted into English, the chemical structures converted into SMILES, and the whole system fitted out with the usual network of embedded links to guide the user from a given molecular structure to other entries describing its function and the various disorders it was designed to treat. And of course this new database had to be as structure searchable as all the rest of Daylight's product offerings, and so that function had to be programmed in as well.

But all of it was now nearly complete, as Dave illustrated with a set of examples. If you wanted to "fortify the spleen," for instance, which was something that a Chinese physician might well prescribe for a patient, then the TCM software would take you to the Manyprickle Acanthopanax Root, whose entry was crammed with information about the plant, its family, geographical origin, biological functions, and medical indications, among other things. Most important, the entry also included a list of the sixteen chemical compounds that were the root's active ingredients, plus their chemical structures, which of course were shown in full color.

The database was a landmark feat, but the question for many of the Western drug developers in the audience was what exactly it meant to "fortify the spleen," "boost the qi," or treat "yang vacuity," whatever that was, or do any of the other strange-sounding things that were standard practice in Chinese medicine. (The TCM, after release, would not in fact sell well.)

Later that night, dinner was to be at Fuller Lodge in Los Alamos, preceded by a group tour of the Bradbury Science Museum. And so at about four o'clock, immediately following Dave's TCM talk and computer demo, the conference attendees piled into a large bus and some cars for the forty-five-minute ride to the Atomic City.

It was a trip back in time, back to the Info Mesa's point of origin, back to the initial collision of instrumentation, computation, and theory that over the years had indirectly spawned every last informatics company in Santa Fe and its environs. The group, including Dave and Dawn, Yosi Taitz, Anthony Nicholls of OpenEye, Bill Ellis of the Walter Reed Army Institute of Research, Dick Cramer of Tripos, the man who had 26 trillion virtual molecules banging around in his personal computer and who had moved to Santa Fe in part to be near Daylight and Dave Weininger—they and all the rest of the crew filed through the Bradbury Science Museum and silently beheld some of the last surviving artifacts of the Manhattan Project, the iconic effort that had started it all.

They saw the old wooden swivel chair that Robert Oppenheimer had regularly used while he had been scientific director of the bomb project, and which he'd sat in for the last time on a commemorative visit to the lab in the 1960s, shortly before his death. The chair was now housed in a sealed glass display case.

They saw the original Los Alamos file card on Nick Metropolis, the physicist assigned to work with Edward Teller, who would wind up as the lab's computer guru and who had played a fundamental role in solving the "Los Alamos problem." The card had been typed up by Dorothy McKibben in the lab's Santa Fe office at 109 East Palace Avenue, on the very day Metropolis first arrived in the city, April 12, 1943. (Metropolis had died in New Mexico on October 24, 1999, at the age of eighty-four.)

And they saw two bombs: an exact copy of Little Boy, the training model of the uranium-235 bomb, identical to the one dropped on Hiroshima; and an exact copy of Fat Man, the training model of the plutonium-239 bomb, identical to the one dropped on Nagasaki.

All of which was sobering in its way. Still, the banquet at Fuller Lodge, the old Los Alamos Ranch School dining hall built of heavy hand-hewn wood beams, was a riotous affair, complete with wine, beer, jazz band, and dancing, although perhaps not as wild as many of the parties held there by the atomic scientists themselves during the 1940s, as they laid bare and converted to practical, usable, orderly information the innermost secrets of matter.

NOT ALL INFO Mesa companies had survived the business recession or the September 11 attacks as well as Daylight had. By the end of 2001 the BiosGroup had laid off some staff members, as had the National Center for Genomc Resources. Prowess Software, Stu Kauffman's Bios spin-off, gave up its Harkle Street offices and returned to its mother company in the Bios building, where it operated with reduced personnel. Other companies continued on more or less as if nothing had happened, while one or two Info Mesa firms had actually prospered in the post–September 11 universe.

QTL Biosystems, located in a business park on the west side of Santa Fe, had been developing biosensing technology that could detect the presence of specific biological molecules, compounds, or even whole organisms such as viruses and bacteria, including anthrax spores. The company's two cofounders, Duncan McBranch and David Whitten, had started the business in 1999 while they were on an "entrepreneurial leave of absence" from the Los Alamos National Laboratory. While still at the lab, they had invented a new type of molecule, a "quencher-tether-ligand," a Rube-Goldberg molecular structure that had the property of causing a fluorescent polymer to light up when in the presence of certain types of small biological objects.

Crux of the invention was the fluorescent polymer, a type of plastic that glowed in the dark. The QTL molecule was a three-part affair, one portion of which, the ligand, had the ability of binding selectively only to the specific type of biological molecule to be detected—a particle of a certain virus, for example, or an anthrax spore. When bound to the spore, the second portion of the structure, the tether, caused the third portion, the quencher, to be physically pulled away from the fluorescent polymer. While it was still connected to the polymer, the quencher interrupted the polymer's natural and spontaneous emission of light, but when the quencher was pulled away by the tether, the polymer once again radiated its characteristic glow.

The effect was not "just like" turning on a light—it *was* in fact turning on a light: the polymer simply shined like a light bulb when the ligand fastened itself to the target molecule, causing the tether to pull the quencher from the polymer. The effect, moreover, was instantaneous: no longer would a diagnostic laboratory have to go through a long and laborious series of chemical tests to determine if a virus was present in a given sample; the QTL sensor worked in less than a second, at what the inventors called "QTLightspeed."

When McBranch and Whitten had tested their system against cholera toxin, the QTL molecule proved to be so sensitive that it detected the poison in extremely dilute ("sub-nanomolar") concentrations, meaning that it would be perfect for poison detection in reservoirs. In the wake of the anthrax-letter attacks, the question was whether QTL's technology would be equally sensitive to the presence of anthrax spores.

"We have developed a prototype hand-held fluorometer that can detect various analytes at low concentrations in a few seconds," McBranch said in early 2002. "We are currently working with military partners to adapt this to anthrax, using reagents and testing facilities available through the federal government."

By that point, the U.S. Defense Advanced Research Projects Agency had given the company $4.5 million to perfect the technology. More had come from private investors and the company was strong and growing, with excellent prospects for the future.

STU KAUFFMAN, WHOSE prospects were never dim, was wondering what to do with the rest of his life.

"I seem to be having an episode in my life which is half business, half science, and I'm having a lot of fun doing it," he said. "The question is what I want to do three or four years from now? Do I want to retire, do I want to become a playwright?"

Even he didn't know. He did, however, finally elucidate the mystery of his sudden jump from science to business. Not that it was ever really a mystery, at least not to him.

"I do have an entrepreneurial bent," he said. "I always have had. I mean, based on the Ballivet-Kauffman patent, I started a biotech company."

That had been Darwin Molecular, the firm that he had been forced out of shortly after founding it. Not that this had been much of a loss: the company had long since merged with Celltech Chiroscience and no longer existed as a distinct entity.

"This is the same side that makes me do experiments," he continued. "Most theorists don't do experiments, but I ran an experimental lab for twenty-three years. So I've done experiments and published experimental papers and so on. Very few theorists do that. But I've always wanted to ground myself in fact, and that's why I went to medical school when I wanted to be a philosopher: I wanted to ground myself in reality rather than just do theory."

So if the BiosGroup seemed to represent a fundamental departure for Kauffman, the fact is, it wasn't.

"Partially I thought it would be fun to make some money," he said, "but that wasn't a big motivation. It's a bigger motivation now that it looks like we're succeeding, but at the start it was, *What an interesting adventure!* We're just doing the sciences of complexity. Firms coevolving with one another, trying to make a living, aren't very different from species coevolving with one another, trying to make a living. Why not get out there and find out whether all these bright complexity ideas that me and everybody's been working on actually work in the real world?

"It's a gigantic, bloody experiment!" he said. "That's really the explanation."

ANTHONY RIPPO, WHO never planned on staying with any one of his startup companies for overly extended periods, had long since left Bioreason. In his case, most of the fun of starting companies came at the beginning, after which a certain relatively unexciting routine settled in.

"As a startup CEO my strategy had always been to sell the company when I thought the time was right," he said. For him, sooner was almost always better than later, and in his view the right psychological moment had come in the summer of 2000, after the company had been in existence for only about three years. At that point he worked out a deal with another firm, a venture capital–backed drug-discovery company, Camitro, which would purchase Bioreason for $8 a share.

"It seemed like a great opportunity and win-situation to me, but my board and major shareholders thought otherwise," he said. "Ultimately, the board did not want to give up control of the company and they thought they could do a better job in managing the company and selling it than I could."

Rippo departed the firm with an agreeable severance package in hand, and Susan Bassett became supreme head of Bioreason, which remained an independent company.

Anthony Rippo and his wife thereupon set out on another "adventure," rambling around the country visiting some of their six kids and grandchildren. On returning home he resisted getting into another startup, but his resolve weakened when he heard about two Santa Fe computer software engineers who were in the process of developing a completely novel, automated stock-trading system.

"So far it looks very promising," Rippo said in January 2001. "Almost too good to be true."

Soon he was the business partner.

The program, it turned out, did indeed conceal a few flaws—not that this was an unprecedented circumstance in software development.

"We brought the automated trading program on line with our own money for several weeks and noticed several major bugs," Rippo said in the summer of 2002. "My partners are still working on the fixes and hope to have it up and running by the end of the year."

He got into a second startup, Catalystiks, soon afterward, a consulting business that aimed at improving a CEO's understanding of his or her own employees. "It amazes me how little time is spent in actually getting to know the people," said Rippo. "CEOs don't really know their management until the time of crisis and then it's too late." For those who wanted to be in better emotional and intellectual touch with their executives, Rippo and his partner, psychologist Leona Stucky-Abbott, would administer personality tests, conduct interviews, and develop psychological profiles on selected company officers.

And because Anthony Rippo was never one to let any grass grow under his feet, he also cultivated a few of his other prime interests, one of which was the role of spiritual factors in human well-being. When he was asked to join the board of the Santa Fe Institute of Spirituality of the College of Santa Fe, he accepted and started a program to reintroduce into northern New Mexico the essence of medieval Christianity—the type of grass-roots Catholicism brought to the area and practiced there in the sixteenth century.

"I believe it is responsible for the unique, open, broad-based, encompassing culture that is found here and which is so appealing to outsiders of all religions, sexual preferences, and cultures," he said.

ANTHONY NICHOLLS, EL Presidente of OpenEye Scientific Software, finally acquired the ultimate, can't-do-without symbol of the successful Santa Fe businessman when in the summer of 2002 he bought a luxury car, a bullet-shaped, 225-horsepower, white Audi TT Roadster—a hot, fast, space-age convertible.

"After six years of unfaithful service, my Jeep died, and my crew insisted I get a nice car," said Nicholls. "They thought that if I started looking the part of a successful CEO, I might act like one. Hah!"

BY LATE SUMMER 2002, neither of the two drugs that had looked so promising initially—neither prostratin, discovered by Paul Cox in Samoa in 1984, nor the possible anthrax drug discovered by Bill Ellis at the Walter Reed Army Institute of Research in 2001—had yet proved out. In fact, neither had gotten to the relevant testing stage.

Prostratin had not entered human clinical trials, which perpetually seemed to be about a year away, each and every year. Irl Barefield, the executive director of the AIDS Research Alliance of America, which held rights to the drug, hoped the trials would begin "soon."

Bill Ellis's anthrax drug had fared even worse: it had not even left the storage warehouse in Rockville, Maryland, where it reposed together with Walter Reed's hundreds of thousands of other experimental chemical substances. The compound in question was a potential drug, a chemical, not a virus or bacterium. It was not dangerous, it posed no hazard to anybody, and the only thing it had to do was to travel some thirty miles northwest, in the back of an unmarked van, to the army's biosafety laboratories at Fort Detrick in Frederick, Maryland, where it was to be tested on animals.

Before such a move could be made, implacable army regulations required that the parties involved complete a Materials Transfer Agreement, which had to be submitted up through channels, signed and stamped by the pertinent officialdom, and received back by those who had requested it. But after more than six months of waiting for this blessed event to occur, the transfer agreement had still not made its appearance.

"The drug is lost in the bureaucracy," said Henry Heine, the Fort Detrick scientist who was to do the testing.

From all of which it could be gathered that however much science and technology might speed up the drug-discovery process, government regulations and procedures remained an impenetrable wall of bureaucratic inertia that could be breached only slowly.

I N M A Y 2 0 0 2, the Los Alamos National Laboratory formally opened the Nicholas C. Metropolis Center for Modeling and Simulation. This was a three-story, 303,000-square-foot structure built to house Q, the lab's new supercomputer. From all external appearances, Q was a throwback to an earlier age of machine computation, for the device was so big and generated so much heat that it was topped by a row of cooling towers that released the thermal waste of computation into the clear New Mexico sky.

Q was rated at 30 teraflops, meaning that it could perform as many as 30 trillion calculations a second, and for the Los Alamos scientists, the machine was the answer to their prayers. It would bring them within sight of the Holy Grail of supercomputing: a full-scale, three-dimensional simulation of a thermonuclear bomb blast.

"Obviously with the various treaties and rules and regulations, we can't set one of these off anymore," said Chris Kemper, deputy leader of the laboratory's computing, communications, and networking division. "In the past we could test in Nevada and see if theory matched reality. Now we have do to it with simulations."

And now they could. This machine would make it all possible; it would do three-dimensional thermonuclear bomb testing "in silico."

Nick Metropolis—not to mention Enrico Fermi, Edward Teller, Stan Ulam, and John von Neumann—would have loved it.

DAVE AND DAWN continued to be the Info Mesa's super-high-tech, high-variety, high-velocity couple. They were into so many projects, doing so many different things so fast, that it was hard to keep up with them.

After the latest MUG meeting, in March 2002, held at La Posada, a group of buildings in a parklike setting just a block away from Daylight's old offices on East Palace Avenue, Dawn took off for Peru for two weeks of mountain climbing. Dave, who took a dim view of getting dirty and was acrophobic to boot, did not go along.

"All that sweating is not my thing," he said.

Soon he left Santa Fe to take a directing course at University of Southern California film school. Like everyone else in the United States, he too wanted to direct, but not feature films or art movies—at least not yet. The fact was that he wanted to improve chemical education on every level, from high school through college and graduate work. Kids loved movies, so why not make movies about chemistry?

Thus was born project CHILE (Chemical Information Lectures and Exercises). As Weininger envisioned it, CHILE would be a series of DVDs that would contain a semester's worth of lectures and labs for use in courses on chemical information or as supplemental material for more conventional chemical course offerings.

"I didn't start out wanting to do anything revolutionary in film," he said. "But I figured that the transfer of intellectual information across generations is so important to society that somebody has to do it for chemistry, and I just wanted to use the most modern technology."

In his films, the starring roles would be played by the legends of chemistry, Nobel Prize winners and the like, people with charisma to spare and who had made fundamental contributions to the science.

Dawn, meanwhile, got involved in a major new project of her own when she and Dave purchased a six-wheel, six-axle surplus Swiss Army truck, a vehicle that could go anywhere, up or down forty-five-degree inclines, across small streams, and over mountain ranges, and converted it into a mobile medical clinic. An incident at the emergency room where she had been working in Albuquerque had caused Dawn to become fed up with the standard insurance-company preapproval and preauthorization requirements for any and all medical care, no matter how urgent. In this case a trauma victim with an intracranial hemorrhage had to be transported across town to another hospital, but hospital-to-hospital transfers, like everything else, had to be preapproved by the insurer. This being a weekend, the insurance company's offices were closed, and the 800 number answered with a recording, so Dawn ended up transferring the patient by taxi.

That was the last straw. She decided to start her own roving medical clinic, a service that would come to the patient, no matter how isolated and far away in the northern New Mexico backcountry. She would accept no insurance whatever and instead would charge every client an upfront, fixed fee of $500.

So Dave and Dawn outfitted the truck, a Pinz 712, loading it up with redundant generators, vaccine refrigerator, electrocardiograph/defibrillator, ultrasound, splints, drugs, a computer system with custom software for controlled drug tracking, computerized medical record keeping, and inventory control, plus a Global Positioning System for offroad navigation.

Dave even helped out with the equipment purchases. "I just bought a used autoclave on eBay for $1,010 which would have been $4,000 new," he said.

And then, finally, as a relief from all this work, Dave made a purchase of his own when he bought a new airplane, his third, a Partenavia P-68 Turbo Observer. This twin-engine, high-wing plane was made in Italy, where it was used mainly for police and government work.

The appeal of the craft for him, however, was that it could land anywhere, even on unprepared desert—there was no need for a runway. The engines were up on the wings out of harm's way, the plane sat high on fixed spring-steel landing gear, and the craft had excellent visibility with windows all around, even in the floor.

Some of Dave and Dawn's friends lived in extremely remote desert locations, places that necessitated a long drive or a short flight, and with the Partenavia they could fly there easily and roll right up to the front door.

One fine, clear morning they decided to visit a colleague of Dawn's who lived on a ranch in the high desert at an elevation of seventy-five hundred feet.

No problem for the Partenavia. Dave pulled the craft out of its hangar at Santa Fe Municipal Airport, he and Dawn climbed in, and they started up and taxied out to the runway. They took off to the north, on a line of ascent that placed the flat plateau of the Info Mesa off the right wing.

Daylight headquarters, with its famed selective muscarinic agonist sculpture sitting just in front of it, was soon visible below.

Then Dave made a slow banking turn to the left and headed west, toward Arizona. Smiling Dave and simpatico Dawn, monarch and medic of the Info Mesa, off together again, into the blue.

SELECTED SOURCES

BIOREASON

Web site:
http://www.bioreason.com

Wrontnowski, Cort. "Counting on Computational Intelligence." *Modern Drug Discovery* (November/December 1999): 46–55.

THE BIOSGROUP, STUART KAUFFMAN

Web sites:
http://www.biosgroup.com
http://www.cistemcorp.com

Kauffman, Stuart. *The Origins of Order: Self-Organization and Selection in Evolution.* New York: Oxford University Press, 1993.

———. *At Home in the Universe: The Search for the Laws of Self-Organization and Complexity.* New York: Oxford University Press, 1995.

———. *Investigations.* New York: Oxford University Press, 2000.

———. "System and method for the synthesis of an economic web and the identification of new market niches." U.S. Patent and Trademark Office. U.S. patent no. 6,125,351 (September 26, 2000).

————. "What's Under the Hood: A Layman's Introduction to the Real Science." Presentation at the Ernst & Young conference "Embracing Complexity 1." San Francisco, Calif., July 17–19, 1996. http://www.cbi.cgey.com/events/pubconf/1996-07-19/proceedings/chapter%2010.pdf

Kauffman, Stuart, and Marc Ballivet. "Method of identifying a stochastically-generated peptide, polypeptide, or protein having ligand binding property and compositions thereof." U.S. Patent and Trademark Office. U.S. patent no. 5,723,323 (March 3, 1998).

————. "Method of identifying cis acting nucleic acid elements." U.S. Patent and Trademark Office. U.S. patent no. 6,100,035 (August 8, 2000).

Meyer, Chris. "Major Thought Leadership Opportunity: Complexity and Stuart Kauffman." Memo to John Avallon, Ernst & Young, June 28, 1995.

COMPLEXITY THEORY

Brockman, John. *The Third Culture.* New York: Simon and Schuster, 1995.

Gell-Mann, Murray. *The Quark and the Jaguar: Adventures in the Simple and the Complex.* New York: W. H. Freeman, 1994.

Gleick, James. *Chaos: Making a New Science.* New York: Penguin, 1987.

Lewin, Roger. *Complexity: Life at the Edge of Chaos.* New York: Macmillan, 1992.

Mackenzie, Dana. "The Science of Surprise." *Discover* 23 (February 2002): 59–62.

Waldrop, M. Mitchell. *Complexity: The Emerging Science at the Edge of Order and Chaos.* New York: Simon and Schuster, 1992.

DAYLIGHT CHEMICAL INFORMATION SYSTEMS, THE SMILES LANGUAGE

Web site:
http://www.daylight.com

Brock, William H. *The Norton History of Chemistry.* New York: W. W. Norton, 1993.

James, Craig A., et al. *Daylight Theory Manual.* Daylight 4.71. 2000. Daylight Web site.

Leo, Albert. "The History of the Development of CLOGP." 1998. Daylight Web site.

Leo, Albert, and David Weininger. *CLOGP Reference Manual.* August 2000. Daylight Web site.

Merlin User Guide. Daylight 4.71. 2000. Daylight Web site.

Thor User Guide. Daylight 4.71. 2000. Daylight Web site.

Weininger, David. "SMILES, a Chemical Language and Information System. 1. Introduction to Methodology and Encoding Rules." *Journal of Chemical Information and Computer Sciences* 28 (1988): 31–36.

———. "SMILES. 3. Depict. Graphical Representation of Chemical Structures." *Journal of Chemical Information and Computer Sciences* 30 (1990): 237–243.

Weininger, David, et al. "SMILES. 2. Algorithm for Generation of Unique SMILES Notation." *Journal of Chemical Information and Computer Sciences* 29 (1989): 97–101.

Wiswesser, William J. "Historic Development of Chemical Notations." *Journal of Chemical Information and Computer Sciences* 25 (1985): 258–263.

DRUG DISCOVERY

Bugg, Charles E., et al. "Drugs by Design." *Scientific American* 277 (December 1993): 92–98.

Cox, Paul Alan, and Michael J. Balick. "The Ethnobotanical Approach to Drug Discovery." *Scientific American* 278 (June 1994): 82–87.

Lemonick, Michael. "Brave New Pharmacy." *Time* 157 (January 15, 2001): 58–67.

Ellman, Jonathan A., and Lorin A. Thompson. "Synthesis and Application of Small Molecule Libraries." *Chemical Reviews* 96 (1996): 555–600.

Plunkett, Matthew J., and Jonathan A. Ellman. "Combinatorial Chemistry and New Drugs." *Scientific American* 281 (April 1997): 68–73.

Werth, Barry. *The Billion-Dollar Molecule: One Company's Quest for the Perfect Drug.* New York: Simon and Schuster, 1994.

H-BOMB, LOS ALAMOS

Aspray, William. *John von Neumann and the Origins of Modern Computing.* Cambridge, Mass.: MIT Press, 1990.

Bretscher, E., et al. "Report of Conference on the Super" (LA-575). Los Alamos: Los Alamos Scientific Laboratory, February 16, 1950.

Harlow, Francis H., and N. Metropolis. "Computing and Computers: Weapons Simulation Leads to the Computer Era." *Los Alamos Science* (Winter/Spring 1983): 132–141.

Metropolis, N. "The Los Alamos Experience, 1943–1954," in Stephen G. Nash (ed.), *A History of Scientific Computing.* Reading, Mass.: Addison-Wesley, 1990.

Metropolis, N., and E. C. Nelson. "Early Computing at Los Alamos." *Annals of the History of Computing* 4 (October 1982): 348–357.

Rhodes, Richard. *The Making of the Atomic Bomb.* New York: Simon and Schuster, 1986.

———. *Dark Sun: The Making of the Hydrogen Bomb.* New York: Simon and Schuster, 1995.

Serber, Robert. *The Los Alamos Primer: The First Lectures on How to Build an Atomic Bomb.* Berkeley: University of California Press, 1992.

Teller, Edward. *Memoirs.* Cambridge, Mass.: Perseus, 2001.

Ulam, S. M. *Adventures of a Mathematician.* New York: Charles Scribner's Sons, 1983.

OPENEYE SCIENTIFIC SOFTWARE

Web site:
http://www.eyesopen.com

Nicholls, Anthony: "Grasp: Graphical Representation and Analysis of Structural Properties." Grasp Web home page: http://honiglab.cpmc.columbia.edu/grasp

PROSTRATIN

Boyd, Michael R., et al. "Antiviral composition." U.S. Patent and Trademark Office. U.S. patent no. 5,599,839 (February 4, 1997).

Cox, Paul Alan. *Nafanua: Saving the Samoan Rain Forest.* New York: W. H. Freeman, 1997.

———. "Ensuring Equitable Benefits: The Falealupo Covenant and the Isolation of Prostratin from a Samoan Medicinal Plant." Draft manuscript, n.d. Institute of Ethnobotany, National Tropical Botanical Garden, Kalaheo, Hawaii.

Cox, Paul Alan, and Michael J. Balick. "The Ethnobotanical Approach to Drug Discovery." *Scientific American* 278 (June 1994): 82–87.

"Prostratin: A Promising Natural Compound for Treating HIV." AIDS Research Alliance *Spotlight* (Summer 2001): 1, 18–20.

SANTA FE INSTITUTE, SANTA FE, THE INFO MESA

Web site:
http://www.santafe.edu

Bass, Thomas A. *The Predictors*. New York: Henry Holt, 1999.

Bulletin of the Santa Fe Institute.

Johnson, George. *Strange Beauty: Murray Gell-Mann and the Revolution in Twentieth-Century Physics*. New York: Knopf, 1999.

Levy, Steven. *Artificial Life*. New York: Vintage, 1993.

Morowitz, Harold J., and Temple Smith (eds.). *Report of the Matrix of Biological Knowledge Workshop*. July 13–August 14, 1987. Santa Fe: Santa Fe Institute, 1987.

Regis, Ed. "Greetings from Info Mesa." *Wired* 8.06. (June 2000): 337–345.

Waldrop, M. Mitchell. *Complexity: The Emerging Science at the Edge of Order and Chaos*. New York: Simon and Schuster, 1992.

ACKNOWLEDGMENTS

FOR THEIR HELP in personal interviews and numerous e-mails, as well as for their kindness in inviting me into their homes, I owe a great debt of thanks to the Info Mesans whose lives, science, and companies I have portrayed in this book: Dave Weininger and Dawn Abriel, Stuart Kauffman, Anthony Nicholls, and Anthony Rippo. For additional assistance I am also indebted to the following individuals.

At Bioreason: Susan Bassett, Bobi den Hartog, David Kuncicky, Sarah Nitzel, and Ruth Nutt.

At the BiosGroup: Kaycee Carroll and Christine McLorrain; as well as Rob Axtell (Brookings Institution), Marc Ballivet (University of Geneva), Anne Crossway (CIStem Molecular), Michelle Hoeft (Presence), Mary Karlton (Prowess Software), Chris Meyer (Ernst & Young), and Laurence Wood (Sunyata Systems). I would also like to acknowledge the substantial help of Mitch Waldrop's excellent book, *Complexity*.

At Daylight: Ragu Bharadwaj, John Bradshaw, Yosi Taitz, Jeremy Yang; as well as Richard Cramer (Tripos), Al Leo (Pomona College), and W. Val Metanomski (Chemical Abstracts Service). Particular thanks to Bill Milne (Ashgate Publishing), for bibliographical help, insider information, and personal favors.

Concerning the Matrix of Biological Knowledge Workshop, many thanks to Harold J. Morowitz (George Mason University).

In Santa Fe I would especially like to thank George Cowan, George Johnson, and Ginger Richardson (Santa Fe Institute).

For their help concerning the history of the Los Alamos lab and the hydrogen-bomb project, thanks to Harold Agnew, Frank Harlow, Roger Lazarus, Marshall Rosenbluth, and Herbert York.

Thanks as well to Alex Heard, of *Wired* magazine; to my agent, Jean V. Naggar; and to my editor at W. W. Norton, Angela von der Lippe.

For her moral support, sound advice, and consistent encouragement throughout the duration of this project, I owe more than I can say to my wife, Pamela Regis.

INDEX